BREATH|PSYCHE|SOMA:
A Clinical Introduction to Breathwork and Trauma Resolution

By

Kyle Buller

Breath|Psyche|Soma: A Clinical Introduction to Breathwork and Trauma Resolution

Copyright © 2019 by Kyle Buller

All rights reserved. This book or any portion thereof may not be reproduced or used in any manner whatsoever without the express written permission of the publisher except for the use of brief quotations in a book review.

Printed in the United States of America

www.settingsunwellness.com

www.psychedelicstoday.com

This book is dedicated to:

Those who have been part of my journey.

Table of Contents

Preface 1

Introduction 2

Chapter 1: 6

Introduction to Trauma and Breathwork 6

    Introduction Breathwork and Models 8

    A Need for a New Approach 10

Chapter 2: 12

Trauma Theory 12

    Introduction to Trauma and PTSD 12

Chapter 3: 28

Neuroscience, Somatic Psychology, and Trauma. 28

Chapter 4: 38

Exploring Breathwork 38

Chapter 5: 48

Clinical Applications of Breathwork for Trauma Resolution 48

    Discussion 67

    Conclusion 68

    Acknowledgments 81

    About the Author 82

Preface

This book is a synthesis of my graduate degree. This book was originally submitted as a capstone paper for graduation. This capstone paper explores the mind-body-spirit connection in relation to treating trauma in the counseling field. Based on my experiences and work, professionally and personally, with trauma, somatic psychology, and breathwork practices, the paper serves as a template for integrating breathwork into a clinical practice to help support clients suffering from trauma or other severe anxiety disorders. Breathwork is powerful practice that encourages clients to regain control of his or her nervous system and can assist in somatic processing of deep-rooted emotions. The first section of this paper will review the literature of modern trauma and somatic psychology research to provide a foundation of trauma theory. The second half of this paper will explore my synthesis of a breathwork framework that could be utilized in clinical work. The framework consists of utilizing breathwork as a way for emotional regulation, somatic processing, and personal and spiritual development. The goal of this paper is to formulate a foundation for breathwork as an intervention and as breathwork training curriculum for counselors and other mental health professionals.

Introduction

Trauma theory suggests that trauma has a significant role in a person's psychopathology and health and wellbeing. There is a popular reframing of the question "what is wrong with you?" to "what has happened to you?" in trauma-informed approaches to mental health (SAMSA-HRSA, n.d.; Levenson, 2017). This reframing helps to normalize and destigmatize the effects of trauma on the body, mind, and spirit. Changing the question to "what has happened to you?" allows one to view mental health issues related to trauma through a news lens and offers a different approach to treating trauma and anxiety disorders. Kylea Taylor (2003) takes this reframing one step further and suggests that "Trauma is not what happens to us; it is what sticks to us." (p. 136). Taylor's (2003) assertion of "sticks" states that traumatic events can *happen* to an individual, but not all traumas *stick* or stay with the person. When the trauma sticks, the trauma becomes frozen in the body and manifests into problematic issues, such as mental health disorders like posttraumatic stress disorder or psychosomatic disorders and physical illness.

Modern neuroscience suggests that traumatic events can leave long-lasting imprints on the body and mind (van der Kolk, 2014). Trauma does not just impact the brain and mind. Trauma also negatively influences physiological responses in the body such as the nervous system and immune function (van der Kolk, 2014). With the new understanding that trauma has a wide influence on the body and mind, are traditional talk therapies enough to treat the various aspects of trauma disorders? Bessel van der Kolk (2014) suggests that top-down approaches such as talk therapy may not be enough, and that incorporating bottom-up approaches are essential in understanding how to work with trauma in all its complexities fully. Breathwork is one of many important bottom-up techniques that needs more research conducted in the counseling field as it can be a crucial intervention to trauma and anxiety disorders.

Personal Statement

I was first introduced to breathwork in 2010 on a cold Halloween day, I had my first Holotropic Breathwork experience. I was then in my first year of undergraduate studies and had enrolled in a transpersonal psychology program. I wanted to explore the phenomenology of consciousness, near-death experiences, and psychedelic experiences. Burlington College offered a weekend workshop in Holotropic Breathwork (Grof & Grof, 2010). I was deeply interested in the work of Stanislav Grof, so I signed up to experience this technique. I approached the workshop with a healthy amount of skepticism; nevertheless, I had a profound and powerful experience of re-experiencing a traumatic birth memory and near-death experience from my teenage years. After this experience, I felt as if I had the opportunity to process deep-rooted emotions in a new way that I was not able to do before. The experience with breathwork allowed me to process these memories and emotions in a somatic way rather than a mental and intellectual way. Since that day, I have been deeply involved in this work, and it has become a meaningful part of my life's path and professional trajectory.

The concept for this capstone project originates from the training that I have received since 2010 from Lenny and Elizabeth Gibson, Holotropic Breathwork facilitators from southern Vermont. The Gibsons have facilitated Holotropic Breathwork retreats and workshops for 25 years. During that time, Lenny and Elizabeth have formalized their breathwork training program to pass their experience onto students who are interested in breathwork. I, along with several other students, have been helping the Gibsons organize their work to formalize their training program, which is now known as Dreamshadow Transpersonal Breathwork. The Gibsons have recently formed a nonprofit to train breathwork facilitators.

Because I have been assisting with many aspects of the Dreamshadow nonprofit over the past decade, components of these efforts should be credited as additional work beyond this capstone project. Although this paper focuses on a clinical approach to breathwork, an important purpose is to integrate my clinical work and the work that I am doing with Dreamshadow, because they both influence the way I approach my somatic theory and work with clients.

Additionally, this paper can serve as a springboard for developing future research projects. To date, clinical studies and research projects regarding the therapeutic potential of breathwork is limited. Consequently, there is a significant need for such a study to establish whether breathwork can be used as an effective treatment intervention for people with trauma.

I have benefitted tremendously from using breathwork in healing my own trauma, and my breathwork experiences and training have shaped my understanding of trauma generally. For example, my own experiences with breathwork have taught me that working with the body can be important when it comes to dealing with trauma as I felt how trauma lives in my body. Breathwork also has allowed me to regain control over my nervous system by becoming more conscious of my breathing patterns. Over the years, I have incorporated various elements of my breathwork training into my work with clients. Although my training in Transpersonal and Holotropic Breathwork operates outside the scope of clinical practice, I am convinced that it would be extremely valuable to integrate various aspects of breathwork into clinical practice.

One aspect that has been helpful is the "sitting" component of breathwork. Sitting has shaped my non-directive approach when working with people in crisis. For example, a client entered my office at my internship site and began to cry and have a panic attack. I let the client cry without trying to stop her or say things to calm her down. I focused on my breath to attune to the client. As I took slow deep breaths and reminded the client that I was there with her, she began to mirror my breathing. Within a minute, the client completely calmed down and returned to a stabilized state. Another component that is useful is the "intensified breathing" to help move energy in the body. Faster and deeper breathing activates an internal process and can help bring emotional content to the surface to be processed. For example, a client that I saw was feeling stuck in her emotions and did not know how to move forward. I encouraged the client to lie on the floor and begin to breathe deeply with minimal pause between the in and out breath. Afterward, the client stated that she felt more open and realized that she had been breathing shallow throughout the session. The client was more animated throughout the session and shared forgotten memories that came up during the breathing exercise.

From these personal experiences, my hope is to stimulate new research and potentially new treatment interventions. The overall goal of this paper and project is to provide the necessary foundation for future training, research, and practice

Chapter 1:
Introduction to Trauma and Breathwork

Traumatic events can radically alter a person's spiritual, emotional, mental and physical well-being. The devastating effects that trauma has on the brain, mind, body, and immune system is well established. For example, a study surveying the impact of adverse childhood events (ACE) reports that a high ACE score increases the likelihood of certain diseases such as heart disease, lung disease and cancer, and overall poor self-reported health (Felitti et al., 1998). Traumatic events can negatively impact not only the person struggling with the trauma but those around as well such as friends and family members (Van der Kolk, 2014). Research in epigenetics suggests that ancestral trauma can be passed down through generations by genes and gene expression (Sapolsky, 2017), implying that we all have been affected by trauma in some capacity. Many trauma-informed theories place trauma at the core of various mental health or physical health issues (Levenson, 2017). Contemporary somatic research suggests that trauma can play a major role in various psychosomatic disorders because the body is the warehouse of experiences and memories (Van der Kolk, 2014). These memories and experiences can thus manifest into physical sensations.

Trauma that becomes stuck, frozen, or unprocessed can lead to the development of posttraumatic stress disorder (PTSD). Morningstar (2017) states that trauma is "the frozen, undischarged energy held in the body and the mind that is often associated with feelings of defeat and helplessness" (p. 72). Lancaster, Teeters, Gros, and Back (2016) estimate that around 8% of the United States population is affected by PTSD, but not every case of trauma will manifest as PTSD. A person suffering from PTSD may seek out treatment with psychotherapy or psychopharmacological interventions or a combination of both

(Reisman, 2016). Psychotherapy proves to be an effective intervention for PTSD and is a first line approach for treatment (Reisman, 2016). For more severe cases of PTSD, a common practice is to combine psychotherapy and medication (Reisman, 2016). Research shows that medication is effective for some, but psychotherapy is the best intervention (Reisman, 2016). One downside of psychotherapy is the high dropout rate (Niles, Polizzi, Voelkel, Weinstein, Smidt, & Fisher, 2018). This is an issue because clients who drop out may not return to seek out additional services or support.

The most common nonpharmacological interventions commonly practiced when treating PTSD include cognitive behavior therapy (CBT), prolonged exposure therapy (PE), cognitive processing therapy (CPT) and eye movement desensitization and reprocessing (EMDR) (Reisman, 2016). Most of the therapies used to treat PTSD involve a cognitive-based approach, which does not address the problem that trauma affects the limbic system and non-verbal parts of the brain. CBT and PE are considered the most common and empirically-supported therapies (Lancaster et al., 2016). If the body stores traumatic memories and affects bodily functions in various ways, it is important to consider other forms of therapy that include the body as part of treatment. Such approaches could supplement the evidence-based practices to enhance therapeutic outcome. Therefore, PTSD interventions should consider somatic approaches as viable adjunct options.

Furthermore, clients benefit from having a variety of treatment options. One under-researched intervention in the counseling field involves breathwork. Breathing has profound physiological effects on the body and mind and plays a large role in the function of the nervous system. Because PTSD and other anxiety disorders are significantly rooted in the nervous and limbic systems, breathing could play a large role in emotional regulation. Price, Spinazzola, Musicaro, Turner, Suvak, Emerson, and van der Kolk (2017) report that a notable study showed a significant improvement of PTSD symptoms when participants engaged in Kundalini yoga, which incorporates various breathing techniques. Therefore, alternative modalities that are sometimes used as an adjunct to therapy should be explored and researched to advance treatment options further.

Introduction Breathwork and Models

Breathwork is a powerful tool for emotional and nervous system regulation, self-discovery, and healing (Morningstar, 2017). There are many forms and schools of breathwork modalities. The general term, *breathwork,* is used to describe a system of conscious breathing practices. The use of breath and breath practices for healing purposes are noted throughout various cultures. Victoria and Caldwell (2013) state that conscious breathing practices have persisted throughout history and that breathwork is the "royal road to physical, emotional, psychological and spiritual health and well-being" (p. 216). Breathwork is one of the main interventions in body psychotherapy (Caldwell, 1997).

Breathwork practices are used for a variety of different reasons and also have different frameworks and purposes. For example, Holotropic Breathwork™ is a powerful breathing practice that fosters self-discovery for spiritual and personal development that consists of fast, deep, circular breathing that incorporates loud evocative music (Grof & Grof, 2010). Yogic breathing, which consists of slow or fast breathing patterns, can be an exceptional tool for soothing the nervous system and regaining control over the emotional body (Descilo et al., 2010). Simple breath awareness practices and mindfulness-based exercises are used in clinical settings to help with emotional regulation (van der Kolk, 2014).

Most of these approaches operate within different frameworks, but they certainly share some basic characteristics. Some breathing practices are not considered evidence-based because there is not enough large scale empirical-based research to consider the efficacy of these practices. Because research lacks in this area, it is difficult to incorporate them into a traditional counseling setting. As mentioned previously, the self-regulating breathing practices and mindfulness practices are widely studied and used in clinical settings, but there is room for expansion in the realm of understanding the application of activating breath practices that foster self-exploration or somatic releases. Victoria and Caldwell (2013) also suggest that there has been little effort or work to incorporate a unified form into the field of counseling and body psychotherapy.

Very few training programs are available that focus on synthesizing both the clinical and self-exploration applications of breathwork. Jim Morningstar's, Therapeutic Breathwork™, is one of the few programs that this writer is aware of that focuses on integrating the clinical and self-exploration applications. The lack of synthesis is problematic for those clients who might greatly benefit from a powerful activating breathwork experience like Holotropic Breathwork or other forms of somatic breathwork techniques. On the other side, activating breathwork practices may not be best suited for people who suffer from a serious mental health issue or severe PTSD history because of the difficulty in providing an adequate level of care. Retraumatization is also a concern when working with populations that suffer from PTSD and other significant mental health issues. People who have difficulty mastering emotional regulation might be better with mindfulness breathing practices that help to calm the nervous system. Victoria and Caldwell (2013) suggest that different breathing techniques may have a specific set and setting, and body-oriented counselors should be able to assess which breathing technique may or may not be appropriate for the client. Victoria and Caldwell (2013) state:

> Some overarching recommendations from the authors are for body psychotherapists to become better breathers, for body-centered counselors to have specific training in a variety of breathing practices and their clinical applications, for breathwork to match the client and his or her particular physiology and psychology and for a clinical capacity to engage with all four treatment models so that the counselor can switch between multiple approaches and methods in conscious breathing therapy. (p. 225)

A Need for a New Approach

The general practice of breathwork is split between the clinical and the spiritual, and very few programs, if any, bridge that gap. In treating people with a history of trauma, it is especially important to be flexible in one's approach, and treatment, as not every tool and technique is appropriate. Victoria & Caldwell (2013) suggest that counselors should use appropriate breathing techniques based on his or her client's needs. The breathing practices should match the client's physiology and psychology, especially when operating within a clinical setting (Victoria & Caldwell, 2013). Rothschild (2017) also emphasizes the importance of teaching self-regulation and stabilization techniques to trauma clients before opening him or her up to deeper emotions and memories.

For example, if a client is experiencing consistent panic or anxiety attacks, a counselor could start a session off by helping the client become aware of her or his breath and breathing patterns. Breath awareness and mindfulness practices can be a simple yet effective technique to help regulate emotions and calm the nervous system (Egan, Hill, & Foti, 2018). For example, Egan et al. (2018) state that engaging in mindfulness practices is associated with decreased resting activity in the amygdala, which is the emotional processing center of the brain. If a client is emotionally regulated and does not present in a high arousal state, activating breathwork can then be encouraged to help bring emotional content to the surface for processing. The idea is to make sure the client feels grounded and safe in his or her body before exposing her or him to a deeper experience. Rothschild (2017) emphasizes the importance of "pumping the breaks" when working with trauma clients and it is important to respect the emotional and physical state of the client. Therefore, a new framework is needed; a system that honors where the client is at and that can assist him or her through various stages of emotional regulation and somatic processing that is not too overwhelming.

Proposed Framework

Utilizing certain breathing practices, such as faster/activating breathwork, in a safe setting can be a profound and life-changing experience for many. Because this paper is examining trauma and PTSD, the clinical overview will focus on these types of disorders. Let us assume that the client is new and in the early stages of processing trauma.

First, the counselor must build rapport and trust before diving too deep into breathwork practices. The counselor should first focus on creating a safe container or setting for the client and help the client cultivate breath awareness techniques to help ground and get into the body. This initial phase of the therapeutic process might take several sessions to establish the trust and rapport necessary for a safe setting. Only then, can the counselor begin to introduce somatic breathwork practices that allow the client to explore emotions and feelings more deeply. Finally, the counselor can facilitate deeper breathwork practices that foster self-exploration and healing, such as Transpersonal or Holotropic Breathwork. (Transpersonal Breathwork is a breathing practice that stems from Stanislav Grof's *Holotropic Breathwork* and has a similar format and style).

Chapter 2:
Trauma Theory

> "No matter how self-assured we are, in a fraction of a second, our lives can be utterly devastated." (Levine, 2010, p. 3)

This section will introduce trauma theory and describe how it impacts the brain and body generally. It will also discuss how simple grounding breathing techniques can help soothe the nervous system to aid in emotional regulation.

Trauma can touch anyone, anytime, and impact that person's life in a split second. It is impossible to predict what types of effects it will have, but safe to say that the consequences for the body and psyche may well be devastating and long-lasting. The healing process is typically a long journey. Traditional talk therapies may be effective but results and changes typically take a long time to develop.

Introduction to Trauma and PTSD

Definition.

The word *trauma* is the Greek word for *wound* (American Counseling Association, 2011). The Substance Abuse and Mental Health Services Administration (SAMHSA) (2018) describes trauma as:

> An event, series of events, or set of circumstances that is experienced by an individual as physically or emotionally harmful or life-threatening and that has lasting adverse effects on the individual's functioning and mental, physical, social, emotional, or spiritual well-being.

Bloom (1999) highlights a definition of trauma stated by Bessel Van der Kolk, who describes trauma as the response resulting from an inadequate ability to cope, internally and externally, from an external threat in one's environment. Levine (2010) suggests that, "Trauma arise when one's human immobility responses do not resolve" and that the "immobility reaction becomes chronically coupled with fear and other intense negative emotions" (p. 67).

Causes.

Trauma can be a consequence of various events causing distress and an emotional response leading to the inability to cope. Morningstar (2017) separates trauma into two categories: events that are most frequently traumatic; and common or jarring events that can be traumatic under certain circumstances. The most common forms of trauma that can lead to trauma disorders are war, rape, assault, the death of loved ones, abuse (witnessing/experiencing), childhood abuse, and severe injury or illness (Morningstar, 2017).

The second cause of trauma that may or may not lead towards a trauma disorder include illnesses, natural disasters, minor car accidents, medical or dental procedures, birth, abandonment, and loud/sudden sounds such as an explosion (Morningstar, 2017). Levine (2010) states, moreover, that medical or dental procedures can lead to the development of PTSD. Levine (2010) reports that around 52% of orthopedic patients developed PTSD following his or her surgery.

What Happens During a Traumatic Event?

A traumatic event can affect the body and mind in various ways. Some of the core symptoms of trauma include hyperarousal, denial, dissociation, constriction, and feelings of helplessness and hopelessness (Morningstar, 2017). Hyperarousal, one of the main symptoms often seen in clients who experience trauma, manifests as anxiety, worry, increased heart rate, muscular tension, cold sweats, and difficulty in breathing (Morningstar, 2017). In response to trauma, the nervous system constricts the flow of blood to the skin, extremities, and internal organs and directs blood instead to the muscles to help the body fight the potential threat. This response affects muscle tone, posture, and breathing patterns (Morningstar, 2017). Dissociation is the body's way of protecting itself from pain. When trauma becomes too overwhelming for the nervous system, the body becomes paralyzed and immobile (Morningstar, 2016). All these symptoms can be viewed as the fight, flight, and freeze responses of trauma. The body and mind's first reaction is to either fight the threat, flee from it, or completely freeze and become immobile.

Chronic Symptoms of Trauma

After the initial traumatic event has subsided, the negative effects on the body and mind typically persist. Morningstar (2017) highlights some of these common, post-trauma symptoms: flashbacks or intrusive thinking, hypervigilance, nightmares, shaking, mood swings, difficulty sleeping, avoidance behavior, hyperactivity, intense emotions, loss of memory or difficulty remembering, fear of dying, and quick or exaggerated reactions to noise or movements (Morningstar, 2017). If trauma is left untreated during the early stages; chronic symptoms may develop. These can have devastating effects on the body.

Morningstar (2017) observes that chronic symptoms of trauma can manifest as low energy, chronic fatigue syndrome, blunt emotional responses, depression, anxiety, and depersonalization/derealization. Chronic stress on the body and mind can compromise the immune and endocrine systems (Morningstar, 2017). Potential consequences include thyroid dysfunction and an array of psychosomatic disorders such as digestive issues, asthma, neck and back problems, and more. These symptoms can come and go and remain hidden for a long period (Morningstar, 2017). This suggests that when certain issues flare up, they may be psychosomatic manifestations of trauma that have taken years to develop because of chronic stress on the body.

Trauma and the Brain

Trauma alters the brain, especially the emotional center of the brain, the limbic system, and memory center, the hippocampus. Understanding how stress and trauma impacts brain function can help counselors and other health professionals treat trauma more effectively. Over the years, neuroscience-based studies have used neuroimaging technology to see the physiological impact of trauma on the brain. The section, *Somatic Neuroscience*, provides a detailed overview of the neuroscience of the brain. The following paragraphs generally outline the different regions of the brain and the function of each.

Limbic system.

The limbic system consists of the amygdala, thalamus, hypothalamus, and hippocampus. It is responsible for processing emotion, behaviors such as fear and anxiety, and different types of memory and attachments. For example, the limbic system becomes activated when a person senses whether situations are either bad or good and safe or dangerous (Siegel, 2011). The limbic system is known as the emotional center of the brain and is also part of the old mammalian brain, which developed roughly 200 million years ago (Siegel, 2011).

PTSD and trauma affect the limbic system in various ways. The recent neuroimaging report by Seo, Rabinowitz, Douglas, and Sinha (2019) shows that chronic stress impacts the limbic system, specifically the amygdala and the hippocampus. These alterations can impact memory, fear and emotional responses, and behaviors (Seo et al., 2019). Many people who suffer from PTSD or trauma have issues with memory or become overreactive and hypervigilant in certain situations (van der Kolk, 2014)

Amygdala and emotional responses.

The amygdala plays a role in processing fear, early memories, attachment, and other emotions across one's lifespan, but it is mostly associated with fear response (Cozolino, 2014).

15

Seo et al. (2019) state that increased activity in the amygdala is associated with emotional reactivity, such as negative emotions and fear. The amygdala is the region of the brain that becomes overactivated and is hyper-responsive to environmental factors in those suffering from PTSD, which can lead to high arousal states and exaggerated emotional/fear responses (Li, Hou, Wei, Du, Zhang, Liu, and Qiu, 2017). Specifically, these researchers report that increased activity in the right part of the amygdala of trauma victims was associated with emotional distress.

When the amygdala becomes activated, it can bring traumatic memories into awareness and elicit an emotional response, such as a sense of danger. Cozolino (2014) suggests that the amygdala can process social and environmental information faster than a blink of an eye. People suffering from PTSD can become extremely sensitive to social cues and facial expressions, which can cause trauma survivors not to trust people or the environment easily. Lacking trust in people and the environment can cause problems in the therapeutic alliance in therapy. Recent research with 3,4-methylenedioxy-N-methylamphetamine (MDMA) shows that MDMA decreases activity in the left part of the amygdala, which reduces stress response and enables the client to establish rapport with a counselor because he or she may not become as fearful or reactive to negative facial expressions (Mithoefer, 2017). The ability to cultivate trust inside the therapy room can thus ripple out into the client's day-to-day life, in the hope that the client can begin to cultivate trusting relationships.

Hippocampus and memory.

Described as the master "puzzle-piece-assembler," the hippocampus is involved in assembling emotional and perceptual memories into biographical memories and also plays a role in organizing explicit memories (Siegel, 2011). Cortina and Liotti (2007) state that the primary function of the hippocampus is to process and consolidate implicit memories into explicit memories. The consolidation of implicit memories into explicit memories allows a person to reflect on the experience. Siegel states, "We have to assemble these implicit puzzle pieces into explicit form in order to be able to reflect on their impact on our lives" (p. 155).

Memory is an extremely important aspect of traumatic experience and recovery. According to Crawford (2010), people who suffer from PTSD tend to "re-experience" the past by "remembering" in the present moment. It is as if the past is happening in the now, and this is the reason for exaggerated emotional responses. Sometimes these "re-experiencing" moments are called, flashbacks. Siegel (2011) suggests that flashbacks may be an activated implicit memory that does not consolidate into an explicit memory. Implicit memories are learned behaviors, which helps explain why, when the memory becomes activated, the person behaves as if the event is happening in the present moment.

During times of extreme stress, the body releases a stress hormone called cortisol. High levels of cortisol can reduce activity in the hippocampus, which in turn can block explicit memories from forming (Siegel, 2011). Often people who suffer a traumatic experience will forget it and then struggle to remember what happened. Cortina and Liotti (2007) explain that trauma can cause dissociation, which leaves the trauma survivor with a fragmented memory of the event. Fragmented memories create an additional consequence of confusion for a person suffering from trauma, often causing complete uncertainty as to whether the experience even happened at all. The implicit memories, however, continue to live in the body and mind. Crawford (2010) states that the fragmented memories recollect as a bodily or sensory memory.

After a traumatic event, a person may keep suffering from memory issues because chronic stress continues to impair the hippocampus (Seo et al., 2019). Roncada, Vandevelde, and Calsius (2018) report animal studies showing that trauma resulted in a decrease of cortisol receptors in the brain. When one has difficulty integrating or resolving her or his traumatic experience, the state of chronic stress can become a sense of normalcy. Morningstar (2017) states, "When appropriate action cannot be taken to return to the body to safety, these emergency states become the common mode of functioning" (p. 77).

How does Trauma Affect the Body?

As shown, trauma affects the brain in many ways. It also affects many other parts of the body. Trauma has an immediate impact on the sympathetic and parasympathetic nervous systems. Consequently, when a traumatic event occurs, the body responds in various ways. The bodily responses are referred to as the "fight, flight, and freeze" responses. Levine (2010) adds two more responses: "The A and Four F's." These are: arrest, flight, fight, freeze, and fold. Arrest happens when a person perceives danger or threat. The brain and body become more vigilant and begins to scan for the danger or threat. If the threat or sense of danger is confirmed, the body will naturally try to escape (flight), fight, or freeze. When body and brain become extremely overwhelmed, the body becomes immobile and cannot move, or it collapses in overwhelm (Levine, 2010). A person who survives the threat or danger breaks free from the sense of immobilization, escapes and then feels empowered to have survived. Levine (2010) states, "Trauma occurs when we are intensely frightened and are either physically restrained or perceived that we are trapped. We freeze in paralysis and/or collapse in overwhelming helplessness (p. 48). Levine (2010) continues, "This collapse, defeat and loss of the will to live are at the very core of deep trauma" (p. 49).

That body's immobility mechanism serves as a survival function (Levine, 2010). In the animal world, freezing or playing dead can offer a form of "invisibility" from a predator. Also, the passivity of an animal can deflect a predator attack (Levine, 2010). Levine also suggests that in the animal world, if one animal plays dead or becomes immobile, the survival of the group or whole is much greater. Trauma responses are essentially survival responses. Trauma becomes an issue when the body continues to live a state of immobility or disassociation. During the collapse phase of trauma, when the body becomes immobile and loses motor function, the body floods the system with endorphins, which act on the pain system to dull any potential pain. Those who continue to suffer from trauma typically continue to feel numbing.

The vortex of trauma.

Trauma can have lasting effects, and healing from it can be a long process. Traumatized individuals can become stuck in what Levine calls the "vortex of trauma," a downward spiral of feedback loops characterized by a cycle of becoming fearful and then experiencing paralysis (Levine, 2010, p. 68). Therefore, trauma can be so difficult to treat. The affected person can become stuck in a state of fear regarding her or his internal state and bodily sensations (Levine, 2010). The fear that arises within the body from experiencing these sensations can then trigger a state of paralysis as the person becomes overwhelmed with emotions (Levine, 2010). The fear does not just revolve around bodily sensations or internal states but deepens due to the state of paralysis. Levine (2010) mentions that the fear of paralysis triggers more fear, which deepens the paralysis even more.

Understanding this cycle is important for effective treatment. Trauma does not just live in the mind, but also in the body. Levine (2010) suggests that clinicians should learn how to uncouple the experience of fear with the experience of immobility and paralysis. Levine (2010 states, "...the very key to resolving trauma is being able to *uncouple and separate the fear from the immobility*" (p. 56).

The nervous system and trauma.

The nervous system includes: (1) the central nervous system, which consists of the brain and spinal cord and (2) the peripheral nervous system which refers to the nerves outside of the brain and spinal cord. The peripheral nervous system consists of the autonomic and somatic nervous systems, which send information from the sense organs, glands, and the rest of the body to the brain for processing (Cozolino, 2014).

The autonomic nervous system (ANS) is typically discussed in the context of trauma because the ANS is associated with self-regulation and adaptability (Agorastos, Boel, Heppner, Moeller-Bertram, Haji, Motazedi, Yanagi, Baker, & Stiedl, 2013). Williamson, Porges, Lamb, and Porges (2015) state that the ANS plays a crucial role in all physiological and behavioral functions that rely on cardiac, smooth, and striated muscles. This explains physiologically why, when trauma strikes, the body freezes, attempts to escape or becomes immobile. The ANS has also evolved to process information related to sexual arousal, courting, and establishing social bonds (Porges, 1998).

The ANS consists of two parts, the parasympathetic nervous system (PNS) and the sympathetic nervous system (SNS). The sympathetic nervous system is an excitatory system responsible for increasing energy for fight or flight behaviors and impacts heart rate, sweat glands, and blood pressure (Porges, 1998). The parasympathetic nervous system is a calming system that helps to dampen the excitatory nervous system and is associated with rest, energy conservation, digestion, and lowering the heart rate (Meg-Rottweil, 2016).

Trauma affects the sympathetic and parasympathetic nervous systems. People suffering from PTSD oscillate between fight or flight behaviors and withdrawal or dissociation behaviors (Williamson et al., 2015). They have difficulty entering a restful state between oscillations and find it difficult to form positive social attachments. The activation of the sympathetic nervous system and dampening of the parasympathetic nervous system can create a large imbalance in the body. Williamson et al. (2015) report that the changes in the ANS can negatively impact immune and endocrine function, which can lead to cardiovascular disease and possible brain dysfunction. Gupta (2013) suggests that this complex interplay between the sympathetic and parasympathetic nervous systems can result in a wide range of somatic and psychosomatic issues, which in turn can contribute to unexplained or undiagnosable medical issues.

Polyvagal Theory.

The complex interplay between the sympathetic and parasympathetic nervous system is addressed more in-depth by Stephen Porges' Polyvagal Theory. The vagus nerve is the tenth cranial nerve that runs from the brainstem to multiple points in the body (Porges, 2003). The name *vagus* (wandering) is apt, as the vagus nerve wanders down the brainstem to the throat, heart, lungs, and digestive system. The vagus nerve is important for respiratory function, heart rate, digestion (Akdemir, 2016), and social engagement/communication (Cozolino, 2014). Recent neuroscience and somatic research emphasize the importance of the vagus nerve in healing from trauma. Cozolino (2014) says that the vagus provides "rapid continuous feedback between the brain and body to promote homeostatic regulation and the optimal maintenance of physical health and well-being" (p. 54).

The polyvagal theory was developed by Dr. Stephen Porges to explain how autonomic function relates to and influences behavior. Porges (2007) states "primary emotions are related to autonomic function" (p. 6). This points to how bodily reactions and emotional reactions are interconnected, and why focusing on bodily experiences is important when treating trauma. For example, Porges (2007) explains that fight or flight responses and states of high arousal are associated with a decrease in vagal influence while an increase in vagal influence may foster social engagement and safety.

Gupta (2013) describes Porges's polyvagal theory as a "complex interplay between the parasympathetic and sympathetic nervous system in PTSD with three hierarchically organized responses" (p. 89). These include the sympathetic fight, flight, and freeze response, immobile and collapse parasympathetic response, and the social engagement response (Gupta, 2013). Williamson et al. (2015) explain that the name *polyvagal* describes the two vagal circuits in the body. The first circuit is associated with the ancient mechanism responsible for the fight, flight, and freeze responses (Williamson et al., 2015). This circuit consists of two parts: the first is associated with the sympathetic nervous system (Cozolino, 2014), which is associated with mobilization and with defense (fight or flight system) (Williamson et al., 2015). The second part of this vagal circuit is the parasympathetic response known as the vegetative vagus (Cozolino, 2014) and is associated with immobilization and shutdown (Cozolino, 2014). The dorsal vagal complex mediates this shutdown response (Gupta, 2013). These ancient defenses have evolved for survival (Porges, 2007).

Smart vagus and social engagement.

The second circuit described by polyvagal theory involves a more recently developed aspect associated with safety, rest, social behavior, and social engagement (Williamson et al., 2015). This is known as the smart vagus or the social engagement system (Cozolino, 2014). The ventral vagal complex (Gupta, 2013) mediates the myelinated smart vagus (Porges, 1998), which is important for regulating heart rate and for engaging in social connections (Porges, 1998). Cozolino (2014) states that the smart vagus is crucial for emotional attunement, caregiving, and sustained social contact.

Trauma can negatively impact interpersonal relationships for a variety of reasons; one of the most important being a compromised felt sense of safety. The smart vagus influences the facial muscles, including the eyes, mouth, face, and inner ears and plays a role in communication. Cozolino (2014) states that since the smart vagus connects the mouth, throat, and stomach and has such an essential role in communication, it facilitates the so-called "gut reaction," experience that occurs, without words, when people interact. This is very significant if a person has poor vagal tone because the person's response is more likely to come from the past, rather than the present experience. Research shows that children who suffer abuse or trauma early in life may be especially sensitive to negative facial expressions (Masten et al., 2008). People who suffer from trauma also have difficulty reading and processing positive non-verbal facial and body language. This can create problems with trust during treatment because the client is more likely to become fearful of the counselor. Cozolino (2014) states:

Breathwork and Trauma Resolution

Traumatized clients are little able to receive support and caring from their well-intentioned counselors – not because they don't want to, but because they are stuck in a primitive root of immobility with its greatly reduced capability for reading faces, bodies and emotions; they become cut off from the human race. (p. 111)

A thorough understanding of the vagus system is necessary when helping clients manage reactivity, cultivate resilience, and ultimately move towards recovery (Williamson et al., 2015). Cozolino (2014) describes the smart vagus as a volume control for the emotions and says that positive vagal tone can increase sustained social engagement and interactions, improve physical health, and helps strengthen the ego. Social withdrawal and isolation are symptoms of PTSD, and stronger social engagement is crucial for health, well-being, and even basic survival. Any understanding of vagal tone is incomplete without considering the role of breath. As Porges states, "breath is extremely important because it turns on and off myelinated vagus," (Joy96815, 2013) which is essential for strengthening vagal tone and regulating emotions.

Bessel van der Kolk (2014) states that Porges's Polyvagal theory provides a biological understanding of safety and danger, which include a complex interplay of bodily sensations influences by the vagus nerve and social interactions, such as observing nonverbal expressions in another persons' face or hearing the tone of a person's voice. Overall, Porges's theory brings social relationships to the forefront of understanding trauma and provides new interventions for strengthening the arousal and emotional regulation (van der Kolk, 2014).

Introduction to PTSD

What is PTSD?

Posttraumatic stress disorder (PTSD) is a psychiatric disorder that was introduced as a diagnosis in 1980 in the Diagnostic and Statistical Manual of Mental Disorders (DSM) (Crawford, 2010). The diagnosis of PTSD has been updated and revised in the current DSM-V and is now renamed under

"Trauma- and stressor-related disorders" section of the DSM-V (Roncada et al., 2018). To meet the criteria of PTSD, a person must be experiencing symptoms such as intrusive thoughts, avoidance or numbing, hyperarousal responses, or significant stress (Zalaquett, 2013) for one month or longer (American Psychiatric Association, 2013). The DSM-V has included "Acute stress disorder" under the "Trauma- and stressor-related disorders," which states that symptoms are experienced from one to three days after the event but can last up to one month (American Psychiatric Association, 2013).

When trauma is not processed and continues to cause significant distress in a person's life, it can then present as the clinical diagnosis of trauma disorder such as PTSD. Trauma can affect the body and mind in many ways, but ultimately it impacts the nervous system and stress response system. Common feelings or reactions of trauma include guilt, shame, disconnection, rage, and isolation (Mental Health America, n.d.). A person who has experienced trauma can also feel in a state of constant alert, a sense of being on edge for his or her safety and may feel powerless and hopeless (Mental Health America, n.d.). It is important to note that not all traumatic events lead to the development of PTSD. For example, witnessing an injury or death, experiencing an accident, or natural disaster have lower rates of PTSD than a person who suffered abuse, war, childhood neglect or sexual assault/abuse (Zalaquett, 2013).

Demographics affected.

The American Psychiatric Association (2018) estimates that around one in 11 adults will develop PTSD in his or her lifetime. This breaks down to around 3.5 % of the American population (American Psychiatric Association, 2018). Haller et al. (2016) estimate that around 8.3% of the general population will develop PTSD over his or her lifetime. Reisman (2016) reports that PTSD affects around eight million American adults each year. Women are two times more likely to develop PTSD than men, according to the American Psychiatric Association (2018). Lancaster et al. (2016) suggest that there are certain populations that are at a higher risk of PTSD, such as people of marginalized communities who have disadvantages in intellectual, social, and educational status, female gender, people who may have experienced early childhood trauma, and those who may have a family history of mental health problems.

The US military veteran population is notable in suffering greatly from trauma disorders and PTSD (Reisman, 2016). Veterans are at a much higher risk of developing PTSD than the general population, and also experience a lack of adequate treatment (Reisman, 2016). Haller et al. (2016) estimate that 18.7% to 30.9% of veterans diagnosed with PTSD are Vietnam veterans, and around 21.8% are returning from Iraq and Afghanistan. Reisman (2016) suggests even higher numbers: the percentage of veterans from Iraq and Afghanistan could be as much as 30%. Reisman (2016) reports that around 500,000 U.S. troops who served in a war are diagnosed with PTSD.

The cost of trauma/PTSD.

PTSD can lead to an array of issues, such as divorce, substance abuse, homelessness, unemployment, co-occurring disorders, drug and substance abuse, criminal activity, chronic health issues, and even suicide/death (Koven & Steven, 2018). The cost of trauma can have detrimental effects not only for an individual suffering from the disorder but also on friends, families, and communities. Research of adverse childhood experiences (ACE) points to some of the costs and other effects of trauma. ACEs can occur when a child experiences trauma, violence, or is a witness of abuse/violence. Stevens (2017) states that high ACE scores raise the risk of chronic disease and mental health issues later in life. For instance, people with a high ACE score are doubly at risk for cancer and heart disease, and at much greater risk for substance abuse (700% higher) and suicide risk (1200%) Stevens (2017). Roncada et al. (2018) also present research that shows convincing evidence about the correlations between childhood trauma and ACEs and the development of many somatic disorders, such as irritable bowel syndrome, chronic fatigue syndrome, fibromyalgia, chronic generalized pain syndrome, and temporomandibular syndrome.

The long-term effects of trauma can be costly on many levels, economically, emotionally, physically, psychologically, and societally. Researching and examining effective treatments is important when it comes to long-term outcomes. Some treatments may be successful for acute symptom management, but how does trauma continue to impact individuals and societies when not fully processed or treated? There is a significant need for research in a wide variety of intervention approaches.

Treatment for PTSD.

The first line of treatment for PTSD and trauma disorders typically involves psychotherapy followed by psychopharmacological interventions. When trauma is severe, a mix of psychotherapy and medications can be helpful. The most researched evidenced-based practices for PTSD include prolonged exposure therapy and cognitive behavioral therapy (CBT) (Reisman, 2016). Of these nondrug treatments, CBT has been demonstrated to be most effective. Prolonged exposure (PE) therapy also has a good outcome record. Reisman (2016) reports that PE is about 60% effective for veterans suffering from PTSD. In this approach, the therapist helps the client revisit his or her trauma in a safe setting while using stress and fear reduction techniques to cope with the trauma (Reisman, 2016). Eye movement desensitization and processing (EMDR) is also another effective nonpharmacological method (Haller et al., 2016).

Psychopharmacological interventions can be an effective form of treatment for some, but Haller et al. (2016) report that psychotherapy is the preferred method for clients seeking treatment. Clients were five to 12 times more likely to choose psychotherapy over medication. One issue with psychotherapy, however, is the client dropout rate, which is reported to be about 18% (Nile et al., 2018). Seppälä et al. (2014) reported that the dropout rate for clients suffering from PTSD can be as high as 54%, and up to 62% for Iraq and Afghanistan veterans. Niles et al. (2018) state that around 13.4% of veterans never showed up to his or her first therapy session in one study and those who did show up, around 22.4% dropped out after the fourth session. One theory about why clients drop out of therapy is that directly dealing with the traumatic material can be extremely difficult, and it is "easier" to avoid the negative emotions or feelings (Niles et al., 2018).

A challenge that often presents in traditional talk therapy occurs when the client revisits old traumas, which can leave the person in a hyper-aroused state and trigger negative emotions. Reisman (2016) suggests that although psychotherapy and other first-line approaches to treatment are helpful, there is also a need to explore better pharmacological interventions for PTSD. In this regard, MDMA-assisted psychotherapy can be especially helpful. This treatment includes both psychotherapy and a medication intervention: MDMA (3,4-methylenedioxy-N-methylamphetamine), also known as Ecstasy. This approach shows significant promise for treating trauma disorders. Clinical studies of the efficacy of MDMA-assisted psychotherapy indicate that the deactivation of the left part of the amygdala (the fear-processing center of the brain) helps clients revisit trauma without being activated and entering a state of high arousal (Feduccia & Mithoefer, 2018). The first clinical trials of MDMA-assisted psychotherapy reported striking results: 85% of the 25 study participants no longer met the criteria for PTSD after the study (Sessa, 2017). Throughout the 16-week study, some participants received three doses of MDMA, while another portion received the control placebo. Even more importantly, the long-term follow up (3-4 years later) found that 85% of the cohort still did not meet the criteria for PTSD (Sessa, 2017). Recent phase 2 studies reported that around 68% of the study participants did not meet the criteria for PTSD after treatment after the intervention (Feduccia & Mithoefer, 2018).

Chapter 3:

Neuroscience, Somatic Psychology, and Trauma

Two fields of psychology, neuroscience and somatic psychology, show special promise for PTSD treatment. Recent results from neuroscience suggest that the brain can heal from trauma. Advanced neuroimaging reveals the brain's capacity to grow and heal via a process called *neuroplasticity*. Somatic psychology research is shedding light on how the body stores trauma and how to properly address this problem in a clinical setting. The combined understanding from these two fields can provide the foundation needed when treating trauma.

Neuroscience

Dan Siegel (2011) says, "One of the key practical lessons of modern neuroscience is that the power to direct our attention has within it the power to shape our brain's firing patterns, as well as the power to shape the architecture of the brain itself" (p. 39). The brain's potential capacity to change and heal over time offers hope for those suffering from PTSD and trauma disorders. Others have reported that the brain can change when the mind changes (Hanson, & Mendius, 2009). When a person directs focused attention towards change and begins to try different or challenging things, the brain's synaptic firings become rewired (Siegel, 2011). Siegel (2011) continues:

> Careful focus of attention amplifies neuroplasticity by stimulating the release of neurochemicals that enhance the structural growth of synaptic linkages among the activated neurons... In sum, experience creates the repeated neural firing that can lead to gene expression, protein production, and changes in both the genetic regulation of neurons and the structural connections in the brain. (p. 42)

New neural wirings form when small changes are made over time, gradually leading to a much larger change in the long run (Hanson and Mendius, 2009). Siegel (2011) describes how meditation is known to create new neural connections and enhance neuroplasticity. Garland and Howard (2009) also stress the role of neuroplasticity in healing the brain from physical or psychological trauma.

Garland and Howard (2009) examined meditation for the potential of its neuroplasticity effects. Focused attention (mental activity) can promote new brain growth, just as physical activity can create new muscle. Imagining or visualizing playing the piano, for example, facilitates new brain growth in the motor cortex. While the research in neuroplasticity and neuroscience is still relatively new, these initial findings have significant implications for the treatment of mental health disorders.

Somatic Psychology

Somatic psychology emphasizes the importance of incorporating the body and bodily experiences into therapy. Susan Aposhyan (2004) states that the first fundamental principle of somatic psychology is that "the body reflects the mind, and the mind reflects the body" (p. 12). The second principle of somatic psychology is that events in a person's early development can continue to influence his or her life, and the body, as the warehouse of emotional memories, is a means of working with these events. (Aposhyan, 2004). Aposhyan (2004) observes, "working through the body provides direct access to early development, nonverbal, and implicit behavioral issues," and that in this way a person can "access the physiological aspects of autonomic neurological regulation, so necessary in the treatment of posttraumatic stress, dissociative processes, depression, and anxiety" (Aposhyan, 2004, p. 13).

The body holds intuitive wisdom and stores emotional wounds. Although many see trauma as a mental disorder, it is also something that happens to the body and persists in the body (Levine, 2010). Cozolino (2014) states that the memories that continue to influence a person are implicit memories: "These experiences are themselves forms of implicit memory; what the mind forgets, the body remembers in the form of fear, pain, and physical illness" (p. 36). Learning how to integrate these implicit memories into a conscious experience is one of the key principals of the therapeutic process (Cozolino, 2014).

Peter Levine (2010) sums up the effectiveness and importance of incorporating somatic practices into therapy:

> Somatic approaches can be enormously useful, or even critical, in this healing effort. They help clients move out of immobility, into sympathetic arousal, through mobilization, into discharge of activation and then finally onward to equilibrium, embodiment and social engagement. (p. 115)

This statement brings us to the important implications that breathwork can have for patients suffering from trauma and PTSD. Breathwork activates and calms both sympathetic and parasympathetic systems and can facilitate a discharge of energy through catharsis.

Somatic Neuroscience

The following infographics provide a quick overview of the brain, brain function, and the nervous system.

SOMATIC NEUROSCIENCE

Introduction to the Brain & The Body

• *Did you know that* •

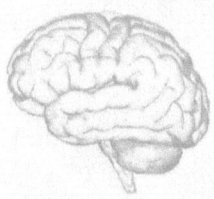 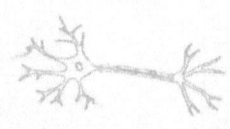

The human brain is a social organ. It is neural network that receives, processes and communicates messages to the environment

The brain is made up of two cells, neurons and glial cells. There are 100 billion neurons that make up the brain

The Brain - interconnected cells that allow electrochemical energy to flow

Electrical energy flows down the axon, converting electrical energy into chemical energy
(Siegel, 2012)

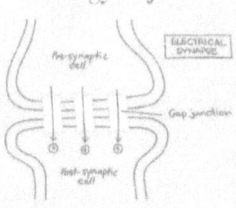

The brain needs relationships in order to evolve and grow

"Relationships are a fundamental and necessary condition for the evolution of the contemporary human brain"
(Cozolino, 2014, p. 6)

THE BRAIN
Parts of the Brain
*Left & Right Hemispheres
Connected by the Corpus Callosum*

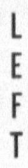  

L E F T

R I G H T

- Linear and methodical
- Past and future-oriented
- Takes in the present moment and picks out details.
- Categorizes and organizes information
- Takes in everything we have learned from the past and projects into the future of possibilities.
- Thinks in language
- "I am."
- Everything is separate

- All about the present moment
- Thinks in pictures
- Learns kinesthetically through movement in the body.
- Takes in all the energy/information around us – understanding what it looks like, smells it, taste like, etc.
- Everything is one.
- Artistic & Creative

(TED, 2008)

Three Main Parts of the Brain
Brainstem, Cerebellum, and Cerebrum

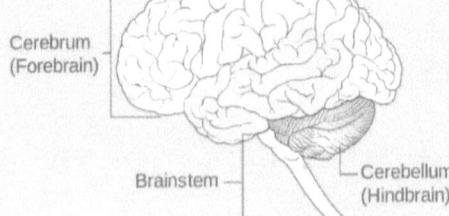

THE BRAIN
Parts of the Brain

The Brainstem (Reptilian Brain)

- Oldest part of the brain
- Controls bodily processes – fright, flight, & freeze
- Controls all vital functions – breathing, heart, sexual arousal, fight or flight, etc.
- Called the "motivational systems"
- Controls our basic needs – shelter, food, safety, reproduction

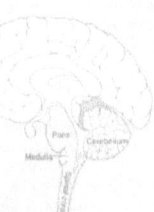

The Cerebellum (Little Brain)

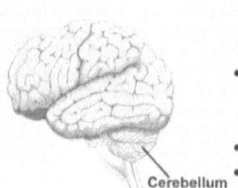

- Contains almost as many neurons as cerebral cortex (Cozolino, 2014)
- Controls motor function
- Controls muscle tone
- Helps with maintaining equilibrium
- Controls balance

The Cerebrum

CEREBRUM CONSISTS OF FOUR LOBES

- FRONTAL
- PARIETAL
- TEMPORAL
- OCCIPITAL

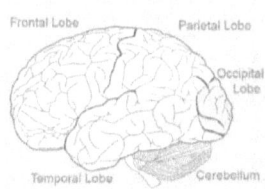

The cerebrum is the largest part of the brain.
The outside gray matter is called the cerebral cortex

THE BRAIN
Parts of the Brain

The Cerebral Cortex (New Mammalian Brain)

- Newest part of the brain and specific to human beings
- It is the outer layer or "bark" of the cerebrum.
- Associated with higher thinking, information processing, and planning.
- The cortex is making maps of your experience (Siegel, 2012)

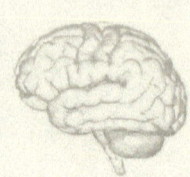

The Lobes of the Cortex

Frontal	Parietal	Temporal
		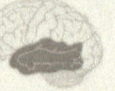
• Regulates: • Motor behavior • Language • Executive functioning • Abstract reasoning	• Links senses with motor abilities • Creates the sense experience of our body	• Auditory processing • Receptive language (understanding language) • Memory functions • Processes visual memories
(Cozolino, 2014)	(Cozolino, 2014)	(Cozolino, 2014)

Occipital

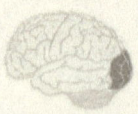

- Responsible for visual processing
- Takes in information from the eyes and processes the visual information

(Cozolino, 2014)

Insular and Cingulate cortices

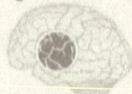

- Integrates somatic experiences for processing
- Gives meaning to somatic/bodily experiences
- Links somatic experiences to cortical and limbic networks for emotional processing

(Cozolino, 2014)

Breathwork and Trauma Resolution

The Limbic System
(Old mammalian brain)

- 200 million years old
- Region of the brain that processes emotions, fear, different types of memory, and secure attachment (Siegel, 2011)
- Our emotional center
- Helps us to form relationships and attachments
- Informs us of what is "good" and "bad" or what is dangerous and safe
- Regulates the hypothalamus (The master endocrine center) (Siegel, 2011)
- Regulates hormones in the body

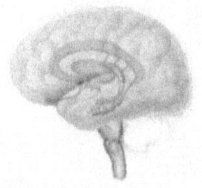

Main structure consists of the amygdala, thalamus, hypothalamus, and hippocampus

Amygdala

- Processes fear, attachment, early memories, and emotion across the lifespan (Cozolino, 2014)
- Can trigger memories of traumatic incidents that inform us of safety and danger
- Mostly associated with fear response.

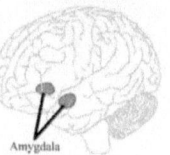

Hippocampus

- Master "puzzle-piece-assembler" that assembles emotional and perceptual memories into biographical memories (Siegel, 2011)
- Plays a role in organizing explicit memories
- Collaborates with the amygdala and cerebral cortex

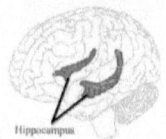

Hypothalamus

- Processes social interactions into bodily processes (Cozolino, 2014)
- Sends and receives hormones via pituitary gland (Siegel, 2011)
- Influences adrenal glands, thyroid, and sex organs

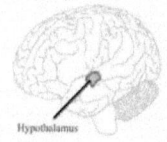

Somatosensory Cortex

Processes information about bodily experiences:

- touch
- temperature
- pain
- joint position
- etc.

Creates our somatic experience

Helps us to make choices by activating implicit memories and can play a role in our "gut feeling." (Cozolino, 2014)

Nervous System

Brain
Spinal cord
Nerves

Central Nervous System (Red)

Peripheral Nervous System (Blue)

Peripheral Nervous System (PNS) consists of the autonomic and somatic nervous systems.

The PNS is involved in sending information of the sense organs, glands, and the body to the brain for processing

(Cozolino, 2014)

References

- Cozolino, L. (2014) *The Neuroscience of Human Relationships: Attachment and the developing social brain*, 2nd ed. New York: WW Norton & Company.
- Siegel, D. (2011). *Mindsight: The new science of personal transformation*. New York, NY: Bantam Books.
- TED. (2008). My stroke of insight | Jill Bolte Taylor [Video file]. Retrieved from https://www.youtube.com/watch?v=UyyjU8fzEYU

Tying It Together: How Does This Relate To Counseling?

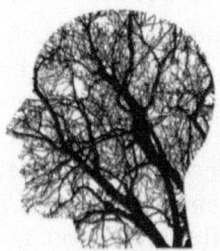

The human brain is complex. It may not be apparent why this information is important for clinicians, but having a basic understanding of the brain can help inform therapists how to work with patients. For example, understanding how the limbic system and amygdala works can help clinicians better serve clients with trauma backgrounds. Understanding how the body and brain work together to create our experience can better inform treatment options as well as potential research in the future.

One area that was not touched on in this infographic is the function of the vagus nerve and parasympathetic/sympathetic nervous system. Learning how to access the vagus nerve via breathing techniques can help calm the body and regulate emotions. As Cozolino (2014) states, the smart vagus acts as "a volume control on our arousal" (p. 86), which can be significant for working with trauma clients or someone who is in a crisis situation.

This information helps clinicians to make better choices when treating clients and can know when to refer out to a doctor or psychiatrist for a higher level of treatment.

References

- Cozolino, L. (2014) *The Neuroscience of Human Relationships: Attachment and the developing social brain*, 2nd ed. New York: WW Norton & Company.

Chapter 4:

Exploring Breathwork

Breathwork is a powerful technique for consideration as a therapeutic tool for clinical use because breathing practices influence the body's sympathetic and parasympathetic nervous system, which can be crucial for regulating and controlling emotional states. *Breathwork* is a term that refers to a wide variety of breathing practices. Crockett, Cashwell, Tangen, Hall, and Young (2016) describe breathwork as:

> A host of spiritual, psychological, and medical practices and interventions that amount to numerous breath-related exercises. A common thread in definitions of the term breathwork is that it exclusively describes breathing that is conscious, intentional, and voluntary as opposed to unattended, reflexive, and autonomic. (p. 11)

Breathwork practices can consist of slow and deep inhales for relaxation or fast and deep inhales for activation. This paper examines a mix of practices as they pertain to trauma.

History of Breathwork

The healing power of breath has been known and applied for millennia. Countless indigenous cultures incorporated breath into their healing practices, which were often based on the belief that psychological and physical ailments manifested from a sick spirit or soul. In many cultures around the world, the word for *breath* and *spirit* were used interchangeably (Crockett et al., 2016). Intentional breathing techniques have also played an important role in early psychotherapy practices.

Etymology.

The word *breath* is used to describe the physical process of inhalation and exhalation, but in many traditions, it also translates to spirit, soul, or life force. In Latin, the word *spiritus* describes the physical intake of air but also is used interchangeably for spirit (Grof, 2000). The Sanskrit word, *prana*, refers to air and breath but also means the "sacred essence of life" in Indian philosophy (Grof, 2000). The Hawaiian word, *ha*, is the word for breath and is related to the word, *mana*, used to describe the spiritual force (Grof, 2000). The word *Chi* is used in Chinese medicine for breath/air is also used to describe the "universal and cosmic energy of life" (Grof, 2000). Similarly, to the Chinese word chi, the Japanese word, *Ki*, describes the life energy and plays an important role in martial arts and other spiritual practices (Grof, 2000). The Greeks used the word, *pneuma*, to describe breath/air and used the word interchangeable for spirit (Grof, 2000).

Traditional breathing practices.

For millennia, cultures worldwide have known and used the power of breath for healing, rites of passage, spiritual connection and other purposes. Anthropological studies confirm that around 90 percent of indigenous cultures around the world held non-ordinary states of consciousness in high regard and recognized their value for religious, spiritual, ritual/ceremonial, and healing purposes (Taylor, 2013). For example, the Essenes practiced an intense ritual for religious/spiritual purposes that involved near-suffocation (Lee & Speier, 1996). Initiates were submerged underwater to elicit an experience of death and rebirth, a ritual considered by many as the original form of (Lee & Speier, 1996).

To understand the role of breathwork in healing ceremonies, consider the Kalahari !Kung Bushmen (or San peoples) of southern Africa. The San participate in various healing rituals that is described as shamanic. The core of these practices involves attaining !kia, a trance state of powerful physical and emotional ecstasy (Lee & Speier, 1996). To attain !kia, the San use rapid and shallow breathing combined with dancing and movement. Those who reach the ecstatic state of !kia can then perform healing rituals on the ill member of the community (Lee & Speier, 1996).

It is impossible here to describe the full spectrum of breathing practices that have been and still are used by indigenous cultures. In some traditions, systems of breathing have developed over time that involves fast or slow breathing patterns and alternating with "locking" or holding the breath in the body, as in certain Indian and yogic traditions. For example, the breathing practice of *pranayama* can include fast-paced, hyperemotional-like breathing patterns as well as holding the breath for extended periods of time and "locking" the breath into a part of the body. Pranayama breathing practices help to cultivate prana, which is life energy. Cultivating life energy is believed to help heal the spirit, body, and mind. Other traditions that utilize breathwork for spiritual, religious, or healing purposes include Sufi practices, Kundalini yoga, Buddhist and Taoist meditations, Tibetan Vajrayana meditation, and Siddha Yoga, to name a few (Grof, 2000).

Re-emergence of breath in modern times.

Breath is deeply valued in countless early cultures, but this connection has been notably absent in modern Western culture, especially in modern medicine, where the potential healing power of breath is largely unacknowledged (Morningstar, 2017). The relatively recent reemergence of breathwork in modern culture for psychological and emotional healing purposes traces back to the work of Wilhelm Reich, who lived in the first half of the 20th century. Reich, a student of Sigmund Freud, is considered the father of somatic psychology. Many of his techniques and theories, including his work with breathwork, have significantly influenced the field of somatic practices (Totton, 2005; Aposhyan, 2004).

Reich 's, *Character Analytic Vegetotherapy,* helped clients loosen what he called *character armor* or muscle tension (Lowen, 1975). It involved various body movements and deep rhythmic breathing practices that fostered emotional and physical catharsis (Lee, K., Speier, P., 1996). Reich believed that the people carried repressed sexual or *orgone* energy in the body

Breathwork and Trauma Resolution

and that it needed to be released. Releasing orgone energy through breathing practices helped clients break through stuck emotions and repressed memories (Lee, K., Speier, P., 1996). Reich's work was adapted and modified by his student, Alexandar Lowen, who developed a technique called *Bioenergetics* that incorporated body movements, exercise, and breathwork. Lowen (1975) stated, "only through breathing deeply and fully can one summon the energy for a more spirited and spiritual life" (p. 66).

During the middle of the 20th century, breathwork gained popularity beyond the clinical psychotherapy setting that began with Reich and moved into the area of personal development and self-discovery. In the 1960s Leonard Orr developed a breathing-based technique, *Rebirthing Breathwork* (Lee, K., Speier, P., 1996), which became a popular modality for healing and self-discovery. Orr's method is based on his own spontaneous rebirthing experience. While in the hot tub, Orr began breathing more deeply and fully, which elicited a profound experience of reliving part of his birth (Lee, K., Speier, P., 1996).

Holotropic Breathwork is another breathing practice that plays an important role in the history of breathwork. Stanislav and Christina Grof developed Holotropic Breathwork in the 1970s and 1980s at the Esalen Institute in Big Sur, California (Grof & Grof, 2010). Holotropic Breathwork is "powerful approach to self-exploration and personal empowerment that relies on our innate inner wisdom and its capacity to move us toward positive transformation and wholeness" (Grof Transpersonal Training, 2018). The practice and technique include five elements; intensified breathing, evocative music, focused bodywork, expressive art, and group process. Holmes, Morris, Clance, and Putney (1996) state that trance states achieved through deep breathing and evocative music help to loosen physical, emotional, and cognitive blocks to allow a re-experiencing of deep-rooted events or experiences in the human psyche. Grof (2000) proposes that such experiences are rooted in three levels of the human psyche; biographical, perinatal, and transpersonal. It is through deep intensified breathing that an individual can re-experience and resolve these issues (Grof and Grof, 2010).

Physiology of Breath – How Does Breathing Affect the Body?

Breath is an autonomous and complex process that affects the body and brain in various ways. Breathing can be involuntary, reflexive, conscious, or behavioral (Crocket et al., 2016). When the body inspires or inhales, the body takes in oxygen and transports the oxygen throughout the body by binding to hemoglobin in the blood (Mithoefer, 2003). Physiological changes in the brain and body occur when voluntarily or consciously changes occur in breathing patterns (Crockett et al., 2016). When one consciously increases her or his air/oxygen intake, oxygen increases in the blood and carbon dioxide is removed (Mithoefer, 2003).

Conscious breathing can create changes in heart rate and vagal tone (Crockett et al., 2016). Various breathing patterns also influence the sympathetic and parasympathetic nervous system (Brown & Gerbarg, 2012). Breath influences many important functions of the body and each breathing style can affect the body and brain differently. More information about the physiology and effects of breathwork are located in the clinical application of breathwork section of the paper.

Hyperventilation.

When talking about breathwork, specifically techniques that consist of fast and rhythmic breathing, there are some concerns about how hyperventilation can affect the brain and body. Hyperventilation is an involuntary response to stress, anxiety, or fear that can cause harm whereas conscious alterations in breath is considered safe. Hyperventilation has many effects on the brain and body. When hyperventilation occurs, the levels of oxygen and carbon dioxide become out of balance (Mithoefer, 2003). Carbon dioxide levels decrease in the blood, which causes an increase in pH levels (Mithoefer, 2003). The changes in carbon dioxide and the increase of pH levels is also known as respiratory alkalosis (Rhinewine & Williams, 2007) and can create the symptoms of hyperventilation, such as lightheadedness, dizziness, and tingling sensations or numbness in the extremities (Rhinewine & Williams, 2007; Mithoefer, 2003). Caldwell and Victoria (2011) report that there has been very little research conducted on the psychological effects of alkalosis.

Breath and emotions.

The limbic system and amygdala are the regions of the brain that are associated with emotions and emotional states (Jerath, Crawford, Barnes, & Harden, 2015). The limbic system and amygdala play a major role in emotional regulation, but Jerath et al. (2015) suggest that one's entire body influences emotions. Jerath et al. (2015) propose that emotions and breath are interconnected, and emotions play a role in breathing patterns and vice versa. Understanding the relationship between respiration and emotional states is crucial for understanding how to treat mental health disorders like anxiety, depression, and trauma disorders (Jerath et al., 2015). Since there is an interconnectedness between emotions and respiration, engaging in breathing practices could help to heal trauma because the breath influences the body and mind in different ways (Brown & Gerbarg, 2012).

Individuals who have experienced trauma may breathe in a shallow and constricted manner. Levine (2014) states that when an individual experiences anxiety, the individual will "stay tense and on guard, feeling fear, terror, and helplessness because our bodies continue to signal danger to our brains" (p. 183). Lowen (1975) also suggests that psychopathology can be observed in the body by observing where a person is holding tension. Lowen (1975) indicates that breathing can help to change states of being and emotions. Caldwell and Victoria (2011) report that observations of children who have asthma also experience negative emotions, which suggests negative emotions can alter breathing patterns. The child may feel her or his airways constricting when they are not physically closing (Caldwell & Victoria, 2011).

Lalande et al. (2012) state that the suppression of negative emotions can lead to a suppression in breathing, which can lead to worry and anxiety. "In humans, conditioned suppression of breathing leads to reduced oxygen and high CO_2 levels in the blood which in turn is associated with a tendency toward increased worry and negative affect" (Lalande et al., 2012, p. 4). The suppression of one's inner experience and emotions can lead to suppressed breathing patterns but can also correlate with "poorer recovery from negative affect, increased sympathetic arousal, and decreased parasympathetic responding" (Lalande et al., 2012, p. 4). Lalande et al. (2012) suggest that once breathing suppresses, the cycle can continue based on biological factors such as increased blood pressure and carbon dioxide levels. Inhibited or suppressed breathing also affect brain metabolism and impacts serotonergic neurotransmission (Lalande et al., 2012). Suppressed breathing can increase the risk of major depression and overall well-being (Lalande et al., 2012).

Breathwork for Trauma

Breathwork and various breathing practices are known to produce feelings of relaxation and promote well-being, which makes breathwork a worthy intervention to explore for mental health issues such as depression and anxiety (Crocket et al., 2016). Brown and Gerbarg (2012) suggest that breathwork may help to treat trauma symptoms and trauma disorders because many of these practices can help an individual discover her or his true self. Physiologically, breathwork practices can activate the parasympathetic nervous system (Brown & Gerbarg, 2012). Activating the parasympathetic nervous system also helps to reduce emotional reactivity (Brown & Gerbarg, 2012), which can be the main symptom of trauma disorders.

The parasympathetic nervous system also plays a role in producing oxytocin, also known as the "love hormone" and helps to facilitate bonding and feelings of love (Brown & Gerbarg, 2012). Individuals who suffer from trauma may not feel safe or trust others easily. The activation of the PNS and the production of oxytocin in the body helps to facilitate human connection and bonding, which leads to feelings of safety (Brown & Gerbarg, 2012). Along with safety, breathwork is also known to activate the vagus nerve and reduce activity in the fear circuits and stress-response systems (Brown & Gerbarg, 2012). The vagus nerve can play a major role in understanding how to treat trauma. According to Levine (2010), understanding the vagus nerve is crucial for knowing how to treat clients with trauma.

The smart vagus, as mentioned earlier, is considered the volume control for emotions. It also plays a role in healthy attachments, bonding, trust, and emotional intelligence (Levine, 2010). Levine (2010) states that therapy usually fails for most clients with trauma because clients who suffer from trauma may feel frightened to develop authentic connections with others, which can lead to isolation. Helping a client strengthen his or her smart vagus and regain control over the parasympathetic and sympathetic nervous systems is important in the treatment of trauma, as strengthening these systems can promote healthier attachment schemas, facilitate bonding, and foster a sense of safety (van der Kolk, 2014).

Many counselors believe that the therapeutic relationship is the cornerstone of therapy. While Levine (2010) agrees that the therapeutic alliance is important, it is not enough when treating trauma clients. "Traumatized individuals are not made whole through the therapeutic relationship alone. Even with the best intentions, and highly developed empathic skills, a counselor often misses the mark here" (Levine, 2010, p. 110). Breathwork is a tool that benefits both the client and the counselor as it helps to strengthen the vagus system, which essential for regulating emotions and aroused states during social interactions (Cozolino, 2014).

Beyond activating the parasympathetic nervous system and vagus system for emotional regulation, relaxation, and stronger social bonding, some forms of breathwork can facilitate non-ordinary states of consciousness that promote healing on a transpersonal and somatic level. Practices like Holotropic

Breathwork or Transpersonal Breathwork can foster deep emotional and physical catharsis, which can aid in healing. Lalande et al. (2012) suggest that the suppression of one's inner experience can inhibit breathing patterns and play a role in depression and anxiety. Breathwork practices can help activate the inner experience and aid in releasing the buildup of energy rather than continuing to suppress it.

Cautions and Limitations of Breathwork

While breathwork does appear to be safe and effective, there are limitations to these practices, and there is a possible risk of re-traumatization. Victoria and Caldwell (2013) state that counselors who incorporate breath practices into her or his practice should use appropriate breathing techniques based on the client's needs. For example, activating breathwork can be psychologically distressing (Victoria & Caldwell, 2013) and may not be appropriate for clients who are not ready to process psychological material. Deep breathing can create anxiety and potentially a crisis for some clients (Victoria & Caldwell, 2013). Victoria and Caldwell (2013) stress the importance of understanding the limitations of each breathwork practice and understanding when these practices can be useful and when they can cause harm.

There may also be a belief in many breathwork circles or practices that breathwork can be a practice for everyone and to trust the process of one's experience. From a clinical perspective, it is important to assess where the client is at and meet the client where he or she is at emotionally and mentally. The American Counseling Association (ACA) (2014) suggests that counselors should not use a technique or intervention if it does not warrant benefit or is potentially harmful. It is important to embody the ethics of "do no harm." The health and safety of the client is the priority when considering interventions and techniques. While breathwork is framed as a practice and technique to help heal trauma or trauma disorders, it is important to remember that there is little clinical data at this point that suggests that breathwork *does* heal mental health disorders.

Breathwork and Trauma Resolution

Breathwork can be used an adjunct to traditional approaches to help heal from mental health disorders.

Chapter 5:

Clinical Applications of Breathwork for Trauma Resolution

Lalande et al. (2012) and others argue that breathwork should be investigated as a treatment intervention within a clinical setting because it is safe and effective. In Europe, breathwork has been gaining recognition as a form of psychotherapy or a practice beneficial if used adjunct to psychotherapy (Lalande et al., 2012). Breathwork has also been gaining recognition in the United States as a healing modality in alternative health movements, but not as much in the realm of counseling. Because breathwork could be used as an adjunct to traditional therapies to help increase the efficacy of therapy, more research is needed to understand its potential benefits.

This section will explore a proposed framework of how to use breath practices in clinical practice. The proposed framework explores this writer's experience with utilizing various breathing and somatic practices. As Victoria and Caldwell (2013) suggest, breath practices should be used appropriately and calibrated to the client's needs.

Models of Breathwork

Young et al. (2010) argue that there should be training for counselors to understand how breathing practices function and to understand the clinical applications for each breathwork practice. Each breathwork practice or technique may or may not be appropriate depending on various factors. Young et al. (2010) suggests that it may be important to have multiple tools in the toolkit and switch between each appropriate practice depending on the clients' physiological and psychological factors.

There are two camps of breathwork; breathing practices that activate and breathing practices that calm and restore the system (Victoria & Caldwell, 2013). Within these camps exist multiple models. Victoria and Caldwell (2013) suggest that the four theories of utilizing breathwork include the relational, energy, regulation, and consciousness models. The relational model describes the importance of breath and breathing patterns in social interactions (Victoria & Caldwell, 2013). This model is related to Stephen Porges's polyvagal theory and describes how breathing can affect our relationships and control our emotions. The energy model consists of breathing techniques that move energy, such as the breathing techniques used by Alexander Lowen (Victoria & Caldwell, 2013). The regulation model describes breathing patterns that are associated with emotional regulation, such as long, deep, and calming breathing practices (Victoria & Caldwell, 2013). Lastly, the consciousness model describes breathwork practices that elicit or foster transpersonal states (Victoria & Caldwell, 2013). These breathing practices could include various yogic practices that help one explore consciousness or Holotropic Breathwork that can foster transpersonal experiences.

Proposed Model: Integrative Breathwork

A model that integrates breathwork into a clinical model for treating trauma should consider incorporating various stages to help titrate the release of trauma. Levine (2010) states that "trauma release must be done in small increments" (p. 82). The idea of titrating the release of trauma is an important insight to incorporate into a breathwork model since some models of breathwork encourage deep catharsis. The model proposed is to help titrate the release of trauma in a clinical setting as the release of trauma can be powerful, overwhelming, and sometimes dangerous or harmful to clients who are not ready to work with powerful and deep material.

Victoria and Caldwell (2013, p. 220) outline suggestions or themes for incorporating breathwork into clinical practice. These themes include:
1. Assessment of the client's breathing patterns and emotional/arousal states to understand whether breathwork could be dysregulating or not.
2. Teach the client how to use breathwork to regulate emotional states.
3. Mirroring breathwork to the client through the

counselor's own breathing rhythm and emotional regulation.
4. Use various breathwork practices to down-regulate or up-regulate a client's emotional state and nervous system.
5. Use the breath as a tool to breathe with feelings rather than suppressing emotions and restricting breathing.
6. Using breathwork practices to create change takes time and should not be rushed.

Incorporating these suggestions, the breathwork model proposed is a system that teaches the client how to use his or her breath to self-regulate the nervous system, to titrate the release of traumatic memories, and to pace the client's healing process so that the counselor does not put the client in a harmful situation such as retraumatization. The model integrates the four models mentioned above, the relational, regulation, energy, and consciousness model, to help "walk" clients through his or her experience in a safer way.

The first stage includes the relational and regulation model as it is focuses on helping a client regulate his or her emotional/arousal state with deep and slow breaths. Using this type of breathwork first helps to educate the client how to use breath for emotional regulation and can help the client and counselor attune to each other. The second stage includes the energy model as it focuses on energy or somatic release. This model focuses on helping the client attune to his or her body and teaching the client to breathe and be with feelings in the body. The second stage incorporates somatic approaches of trauma release. The third stage of breathwork includes the consciousness model and focuses on the healing potential of transpersonal states of consciousness. This stage may or may not be appropriate for a clinical setting considering the duration of the breathwork practice and the context of the practice. Short or modified versions of consciousness models such as Transpersonal Breathwork and Holotropic Breathwork can be incorporated into a clinical setting, but the full practice exists outside of a clinical model.

Stage One: Emotional Regulation and Therapeutic Alliance

Goal.

The first stage of the breathwork model focuses on holding space for trauma in the early phase of therapy. The goal of this stage consists of establishing a good rapport, cultivating a strong therapeutic alliance, and teaching the client how to use slow deep breathing to regulate her or his nervous system. Kim, Roth, and Wollburg (2015) stress that a strong therapeutic alliance leads to change and is one of the cornerstones of therapy. Clients with trauma disorders can experience high arousal states and possibly deal with trust issues early in therapy. The goal is to help the client connect to his or her breath to calm the nervous system and to help establish rapport with the counselor so that the client continues with therapy rather than dropping out. This stage also includes the counselor practicing slow breathing techniques to help mirror healthy emotional regulation.

Theory.

Slow, deep breathing practices can activate the parasympathetic nervous system, which aids in calming the body and mind (Brown & Gerbarg, 2012). The activation of the parasympathetic nervous system balances out an overactive sympathetic nervous system and aids in self-regulation. This can be important early in treatment when a client may experience anxiety or panic attacks. Levine (2010) states, "The capacity for *self-regulation* is what allows us to handle our own states of arousal and our difficult emotions, thus providing the basis for the balance between authentic autonomy and health social engagement" (p. 13). Teaching a client how to self-regulate by using slow deep breathing techniques can help with developing healthy attachment and building a strong therapeutic alliance. Cozolino (2014) states that teaching emotional regulation to clients is not the only goal, but also teaching counselors to cultivate her or his own self-regulation skills.

The role of the counselor is to help provide a safe container for the client to explore their emotions. If a client with a history of trauma or significant attachment issues comes in, his or her brain and body may be attuned to the counselor's own body to figure out if the environment is safe or not. Cozolino (2014) describes that the fusiform face area (FFA) of the occipital lobe is responsible for facial recognition. If a person deems the face unsafe, this could activate the left amygdala (Cozolino, 2014), which could activate a fear or stress response, and create difficulty in the therapy session. A client may not feel like he or she trusts the counselor or may want to leave/drop out of therapy because of not feeling safe.

Facial recognition comes into play when attuning to another person and developing relationships. Cozolino (2014) describes the function of the vagus nerve in social engagement, and how the vagus nerve picks up information from our internal organs and translates it to facial expressions. The vagus nerve influences the facial muscles, which in turn can show others what may be going on internally (Cozolino, 2014). The human brain has evolved and developed a complex social system that can help one to attune to others. Cozolino (2014) speaks about the mirror and resonance systems that can help one attune to the emotional state of others and that this system might have evolved to help humans imitate one another for hunting, gathering and building connections. These mirror and resonance systems "provide us with a visceral-emotional experience of what the other is experiencing, allowing us to know others from the inside out" (Cozolino, 2014, p. 52).

Smart vagus activation through breath.

As previously discussed, the *smart vagus* is one of three systems of Stephen Porges's polyvagal theory. This system, also known as the social engagement system, is a myelinated branch that helps to calm the sympathetic nervous system and sympathetic arousal (Cozolino, 2014). Cozolino (2014) suggests that this system has evolved and developed over time to foster emotional attunement and sustained relationship and also to help enhance caregiving. The smart vagus, as Porges suggests, might have also evolved to help humans communicate among one another without activating a flight/fight response, which would lead to sustained relationships and emotional attunement (Cozolino, 2014).

Why is this important and what are the implications for counseling? As Cozolino (2014) states, the smart vagus acts as the "volume control on our arousal" (p. 86). Unlike the autonomous nervous system, which seems to act more like an on/off switch, and can activate flight/fight responses, the smart vagus helps us to regulate our emotions even in an aroused state and during social interactions (Corolino, 2014). This suggests that we all can control our emotions, even when feeling aroused. Porges mentions that one can use the breath to turn on and off the myelinated vagus and stimulates the parasympathetic nervous system (Joy96815, 2013). It seems reasonable to suggest that a counselor can use this information to inform clients about how to regulate their arousal or emotional state better. A balanced vagus system could help a client develop healthier relationships considering that the smart vagus influences our non-verbal commutation through facial muscles. Deep breathing practices can turn on and off the myelinated vagus and can play an important role in stimulating the parasympathetic nervous system (Joy96815, 2013).

Practices.

Breath Awareness

Most of us are not aware of our breath. It is such an autonomic function, that we breathe without thinking. This practice is just to practice the awareness of one's breath.

- With your eyes open or close, bring your attention to your breath.

- Inhale through your nose, bringing the breath way down into the abdomen. Hold the breath in the belly for a second.

- Exhale through the mouth.

- There is no right or wrong way to do this exercise. The point is to simply bring awareness to your breathing and to begin to breath with a little bit of conscious effort.

- First start off with ten cycles of inhaling, holding, and exhaling.

- This can turn into a deeper/longer meditation/mindfulness practice. Feel free to experiment for a longer period.

Relaxing and Down-Regulation: 4-7-8 Breath

The 4-7-8 breath is a popular breathing technique to help calm and regulate the nervous system. This technique seems to be popularized by Dr. Andrew Weil.

- Begin by exhaling through the mouth completely.
- With your mouth closed, inhale through the nose for four seconds
- Hold your breath for seven seconds
- Exhale through your mouth for eight seconds.
- Repeat this cycle for three to four more times until you feel relaxed.

If this breath practice is activating for any reason, return to your normal breath and take it slow. Deep breathing exercises can make a person feel light or dizzy at times.

Stage 2: Activating Breath and Somatic Therapy

Goal.

Once the client can self-regulate his or her emotional states through deep breathing techniques, the second stage focuses on slowly working with the trauma stored in the body and mind. This stage includes two phases that include various somatic approaches and theories. The first phase works with slow deep breaths but with focus attention on a client's inner experience. This helps the clients to be with his or her inner experience in the *now* to help process trauma. The second phase consists of circular breathing techniques that help to activate the nervous system. Each phase of this stage should be used appropriately based on the client's readiness. The goal of this phase is to help with down-regulation and up-regulation of the nervous system and to process trauma in the body somatically.

Phase one theory: Somatic experiencing.

Processing trauma via verbal expressions may be difficult for some, and sometimes there are no words that can accurately describe the event or experience. Levine (2010) states, "Despite our apparent reliance on elaborate speech, many of our most important exchanges occur simply through the 'unspoken voice' of our body's expressions in the dance of life" (p. 11). Traditional therapies typically neglect bodily experiences and usually focus on top-down processing rather than bottom-up processing (Levine, 2010). The focus of this phase is to create a space to allow the client to experience and process bodily sensations in the *here and now*.

After a client masters self-regulation and is ready to begin processing deeper emotions or trauma, a counselor can begin to guide a client into a somatic experience by including two concepts; being with bodily experiences in the *here and now*, and deep breathing to help stay with the experience or emotions. This phase is about creating a corrective experience without re-traumatizing the client. The way a counselor can facilitate this is to coach the client to turn inward and focus on his or her bodily experience. The counselor can ask the client to locate a feeling in his or her body and to take a couple of deep breaths into the space where he or she feels the emotion or bodily sensation. The goal here is to focus internally on the somatic experience and to "be with" the feeling or sensation without trying to force anything to happen. As a client begins to "be with" her or his experience, sympathetic arousal can happen. Deep breathing can aid in balancing the nervous system to help to return to equilibrium.

Levine (2010) shares the importance of staying with bodily sensations in the here and now. When trauma begins to surface, the body can start to react in various ways, which may include shaking. Levine (2010) states that shaking is a way for the body to restore equilibrium and that counselors should allow shaking or trembling to happen. Levine (2010) mentions that trembling can:

Hold the potential for catalyzing authentic transformation, deep healing and awe… Trembling can help to reset the nervous system and create equilibrium to the body and mind... These gyrations and undulations are ways that our nervous system 'shakes off' the last rousing experience and 'grounds' us in the readiness for the next encounter with danger, lust and life. (p. 16)

The goal during this phase is to help the client develop a relationship with her or his body and bodily experience while building resilience to high arousal states by learning how to breathe and be with the experience rather than suppressing it or pushing it away. When a client can build resilience against arousal states due to trauma, the client can begin healing.

Phase two theory: Activating breathwork.

Activating breathing practices can be beneficial when a client is experiencing states of hypoarousal. When feelings or emotions become stuck or frozen in the body and mind, deeper and faster breathing can help bring emotions to the surface for processing. Activating breathwork is also known as *circular breathwork*, which a client inhales and exhales with a minimal pause in between breaths in a circular pattern. This phase can deepen a client's understanding of her or his emotions as circular breathwork can foster emotional catharsis and can bring memories or feelings to the surface. It is important to stress that when conducting up-regulating breathing practices, it should be done under supervision such as in a counseling session with a trained professional, and the client should have the tools to be able to self-regulate in her or his daily life (Victoria and Caldwell, 2013) since activating breathing practices can bring up deep emotions.

Breathwork theory suggests that bringing deep psychic material to the surface to be processed is integral to healing trauma. Holmes, Morris, Clance, and Putney (1996) state that the "process of working through and integrating the traumatic material often leads to the resolution of seemingly intractable psychological problems" (p. 114 – 115). From a psychodynamic perspective, activating breathwork helps unconscious material to become conscious. Bringing the unconscious to the surface is an important part of one's healing process, but it should be done with careful consideration because re-traumatization can occur.

Morningstar (2017) states that activating breathwork is a tool that helps to process trauma by going directly into the limbic system whereas traditional therapies rely on going through the prefrontal cortex. Activating the limbic system and sympathetic nervous system through breathwork, "Allows formerly implicit-only or partially integrated stimuli to resurface and be more fully integrated with safe outcomes" (Morningstar, 2017, p. 81).

Circular breathing is known to activate the sympathetic nervous system (Nivethitha, Mooventhan, & Manjunath, 2016). Lalande et al. (2012) state that conscious circular breathing is known to change brain wave activity and help to increase alpha brain activity as well as increase the production of serotonin. It is also reported that circular breathwork has been shown to increase dopamine release to around 65% (Lalande et al., 2012). These changes in the brain and biochemistry can lead to positive affect changes such as increased attention, positive mood, motivation, and arousal (Lalande et al., 2012). Lalande et al. (2012) reported that clients who participated in conscious breathwork practices stated an increase in relaxation, mental clarity, and emotional regulation, followed by a decrease in anxiety.

Contraindications and limitations.

Activating and circular breathing exercises can be beneficial and valuable to clients, but there are limitations, contraindications, and clinical considerations (Young et al., 2010). Clients who suffer from asthma, cardiovascular issues such as high blood pressure or heart disease, serious mental illness or hallucinations, and those who have has any recent surgeries should not engage in activating breathwork (Young et al., 2010). If a client does suffer from a serious mental illness, it is important to assess whether the client's mental health has been stabilized for six months or longer as breathwork could potentially trigger negative emotions and exacerbate current issues.

Circular breathing can also produce some unpleasant adverse

effects. A common adverse reaction to circular breathing is carpopedal spasms, also known as tetany (Grof & Grof, 2010). Tetany leads to tightening, and cramping of the muscles, typically in the hands, face, and legs, and can be painful and uncomfortable for many (Grof & Grof, 2010). Other common symptoms of circular breathing include lightheadedness, dizziness, tingling and numbing sensations throughout the body, and an increase or decrease in mood such as fear, anxiety, sadness or frustration (Grof & Grof, 2010; Lalande et al., 2012). It is important to discuss both the positive and negative effects of breathwork with a client and assess whether this practice may or may not be appropriate. Ethically speaking, providing an informed consent about this process is the best practice (ACA, 2014).

Practice.
Here is a breathing exercise that I do at least once a week to help clear pent up energy and emotion in my body. This is a simple breathing exercise and meditation that helps to relieve stress and tension. This exercise also fosters relaxation and a meditative state.

Breathwork to Release Emotional Energy

1. **Create a safe and comfortable space.** I like to dim the lights and turn on my salt lamp. Burning incenses or sage is also helpful to cleanse the room/area of any negative energy. My favorite is Palo Santo! It has such a sweet smell to it.

2. **Create a playlist.** I typically like to create a playlist that ranges from 10-15 minutes. I usually pick out music that is suitable for meditation or yoga. Get the music started before assuming a comfortable position.

Spotify or any other music player works! I just like to find music with drumming or meditative music.

3. **Find a comfortable position.** I like to lay on the ground on my back. You can use a pillow to support your head or use a

pillow under your knees. Use a yoga mat to lay on or you can also use blankets.

4. **Close your eyes** and place both of your hands on your sacral chakra or right below your navel. You can also place your hands-on whatever chakra that you like to, but I typically start by placing my hands below my navel because it helps with guiding my breath.

5. **Bring your attention to your breath and inner world.** Body Scan. As you close your eyes and begin to relax into your body, feel the ground support you, and start to become aware of your breath. Begin to focus your attention on your body and how your body feels as you bring in each breath. Pay attention to any sensations, feelings, or tension in the body.

6. **Focus on deep and slow breathing for 10-15 minutes.** As you ease into your body and listen to the music, begin to practice breathing deeply into your belly. If your hands are on your sacral chakra, feel your breath enter that charka. Feel your belly rise and the sacral chakra activating.

I invite you to inhale deeply and visualize bringing in bright healing energy, holding in your body for a second or two, exhale and release, visualizing a black smoke representative of tension, stress, or toxins. This breathwork exercise is to help clear out any tension in the body.

7. **Let the music guide you.** As you continue breathing deeply into your body, imagine that each breath is bringing in healing energy. If your breathing pattern changes a little bit, do not stress about it. Let any expectations go and just focus on the music and your breath.

8. **Be Curious and Let Go.** With each exhale, imagine that you are getting rid of any thoughts or emotions that your body might be holding onto. Each out breath is a chance to let go even more. Explore the feelings in your body and be curious. What does it feel like to breathe into certain parts of your body or

emotions? What does it feel like to let go and relax into the tension or feeling in your body? What does it feel like to relax into the emotion? If the body wants to shake or let go, allow the body to naturally respond. There is no need to force anything to happen, but just be with the body, its feelings, and allow the body to relax and let go.

9. **Come back to your body.** As the music ends, take your time coming back to your body. As you come back, just pay attention to how you are feeling. Are you feeling different from before? Do you feel lighter? Did any emotions come up for you? Just simply observe any noticeable shifts. When you feel like you are back in your body, slowly open your eyes and sit up. Remember to take it slow!

Stage 3: Beyond the Clinical Setting: Transpersonal Breathwork

Goal.

Breathing techniques such as Holotropic or Transpersonal Breathwork are typically longer-form styles of breathwork offered in a group or retreat setting. These techniques typically operate outside of the clinical setting, but they can offer tremendous benefit to clients who are emotionally regulated and grounded. Brewerton, Eyerman, Cappetta, and Mithoefer's (2012) research suggests that long-form breathwork can be beneficial in residential settings. Brewerton et al. (2012) report that when Holotropic Breathwork was used in adjunct to evidence-based practices, clients reported a decreased in anxiety, depression, and early traumatic memories.

In a clinical counseling setting, these techniques are shortened to 30-45 minutes, but typically the sessions last around three hours. One of the goals of working with this form of breathwork is to help a client process deep emotional material that is rooted not in just the biographical, but the transpersonal and perinatal (Holmes et al., 1996). Holmes et al. (1996) state that transpersonal theory suggests that the "resolution of most forms of psychopathology requires accessing and fully experiencing events rooted in all three levels of the human psyche" (p. 115). As a client moves beyond the biographical, he or she can explore, process, and integrate the deeper layers of the psyche that can foster a deeper healing process.

Theory.

Holmes et al. (1996) states that deeper and faster breathing, accompanied by evocative music, "induces a trance state which facilitates the loosening of any cognitive, emotional, and physical barriers to enter the altered state of consciousness necessary to access the perinatal and transpersonal realms, as well as repressed material from the biographical realm" (pp. 115 – 116).

Utilizing techniques that foster self-discovery and self-exploration can be an important part of the healing process as they can help elicit deeper meaning. Morningstar (2017) states, "The spiritual perspective can help give a broader meaning to the trauma and consequently further catalyze the healing process" (p. 106). Many people who have experiences with Holotropic or Transpersonal Breathwork can have *transpersonal* experiences. These experiences can be *beyond the personal*, as the term transpersonal suggests.

Bray (2013) posits that transpersonal experience and experiences that foster psychospiritual growth can result in greater satisfaction in life, increased creativity, and self-actualization. Grof (2003) suggests that traditional psychotherapy is limited as it primarily focuses on solving solutions or intellectualizing and understanding biographical memories and material. While developing an intellectual understanding of biographical material can be beneficial, Grof (2003) suggests that the resolution of psychopathology can occur when one re-experiences the event and allows the body to process the event or incident somatically. Grof (2003) states that the main goal of working with holotropic states of consciousness is to "activate the unconscious and free the energy bound in

Breathwork and Trauma Resolution

emotional and psychosomatic symptoms, which converts these symptoms into a stream of experience" (p. 76). Activating and intensifying symptoms can thus lead to resolution and healing (Grof, 2003). According to Grof (2003), the symptom of trauma and emotional disorders is a byproduct of the psyche trying to "free itself from old traumatic imprints, heal itself, and simplify its function. They are not only a nuisance and complication of life but also a major opportunity" (p. 74).

Another benefit of utilizing a technique like Holotropic or Transpersonal Breathwork outside of the clinical setting is that these practices provide a group container to aid in the healing process. One of the components of these practices is *group process*, which allows for a client to move beyond the one-on-one session and begin to create a social network. Transpersonal Breathwork offers a community. Taylor (2003) states that breathwork can offer a group ritual experience that "offers one place where we can, together, metabolize grief and trauma" (p. 139) in a community context. Cultivating connections and strong social bonds seems to be an important part of the healing process for those who suffer from trauma and PTSD.

Contraindications and limitations.

The contraindications are the same as listed above in the "activating breathwork" section. Transpersonal Breathwork and Holotropic Breathwork may not be appropriate for those who are in the early stages of processing trauma as emotional and physical catharsis can be retriggering or retraumatizing for some. Although these techniques can foster deep "healing" and self-discovery, this may be limited in some cases. These techniques should not be utilized as a substitution for traditional therapies, but rather should be used an adjunct to traditional treatments. It is important to emphasize that activating breathwork can sometimes intensify symptoms and it is important that those suffering from a serious mental health disorder seek professional help if this occurs. A participant should have professional supports and not just rely on the technique or modality.

Eyerman (2013) observes that there were no adverse effects or reactions in over 11,000 case studies of utilizing Holotropic

Breathwork in a residential treatment facility, but that does not mean that some people are not at risk. In the case of Eyerman's studies, all the participants were in a safe setting with ongoing supervision and support. Such settings are not commonly available for most of the population suffering from PTSD or other anxiety disorders. In more limited cases where a higher level of care is needed, many safety mechanisms are already in place, and breathwork techniques like Holotropic Breathwork may be useful as a treatment intervention.

Components of Holotropic and Transpersonal Breathwork

The components of Holotropic and Transpersonal Breathwork include, Intensified breathing, evocative music, focused body work, group process, and expressive art (Grof, & Grof, 2010; Dreamshadow, 2018).

Intensified Breathing: The breathing pattern is circular; the most important step is to push as much air in and out of the body as possible. These breaths should be deep and continual in a circular process. There is no particular way of breathing, but it is encouraged to use the deep circular breathing pattern. Over time individuals usually find their own breathing pattern once they have induced a holotropic state. Reader claims that it only takes a few minutes of hyperventilation to produce lower levels of carbon dioxide, and these levels can be maintained even with normal breathing (Reader, 1994). It is common to see individuals to start breathing deeply and circular, and once in a holotropic state, their breathing seems to level off and sometimes even slow down tremendously. Hyperventilating is known to produce both theta and delta waves which both are associated with meditative-like states (Reader, 1994).

Closed eyes: To begin with, a person lies on his or her back on a mat with eye shades covering their eyes. The reason for covering the eyes can be seen by the research of Reader, who mentions that, eye closure, darkness, and isolation have been a huge part of most of the world's religions. The reason for this is that physiologically, it is known that total darkness can help induce mystical experiences of the mind (Reader, 1994). Total darkness has been known to enhance visual imagery since Asclepides of Bithynic around 124 B.C.E (Reader, 1994). Reader reveals that the occipital and temporal lobes become hypersensitive during visual deprivation, such as being isolated in darkness or having closed eyes (Reader, 1994).

Visual deprivation can begin to occur just after thirty minutes of total darkness. The physiological effects are that the retina and occipital cortex of the brain begin having impulsive discharges of energy which creates images such as lines, curves, complex geometric structures, and even firework patterns (Reader, 1994). The eyes have always played a major role in many of the world's religions for over the past five thousand years and have always been related to as a mysterious organ which contains the secret mysteries of the world. The eye was the organ that was known to be able to see the "whole truth" by looking for "inner wisdom" (Reader, 1994).

Evocative music: The music is added into the mix to help guide the person's breathing, but music also helps alter consciousness as well. In many traditional cultures, shamans use the drum to enter a trance state. The music acts as a guiding instrument for individuals on their journey. It helps to bring a person to a holotropic state as well as helping with navigating the psyche while in holotropic states. This is very similar to the idea of the sacred plant songs, called, "icaros," which shamans who use "plant medicine" to heal, sing during healing ceremonies to help guide people through their inner world to ensure safety.

Focused Bodywork: Energy, emotions, and memories can be stored inside the body, and if pains arise in certain areas body work is often necessary. The idea is to dive deep into one's self and breathe into the pains while a professional help to apply pressure on the body, releasing the emotional/energy block (Grof, & Grof, 2010). The pressure isn't applied directly by a facilitator, but rather applied by the breather. For example, if a pain develops on a person's back, the participant is asked to lie on his or her stomach while a facilitator places their hands on the spot that is causing pain. The facilitator then tells the breather to "breathe into the pain" and push against the facilitator's body weight. This ensures the safety of the breather and lets them have complete control over the situation. If the pain becomes unbearable, all the breather has to say, is "stop" and then relax. Bodywork can assist in resolving psychosomatic issues in the body.

Group Process: Lenny Gibson states, "We are the descendants of successful tribes." This statement stresses the importance of group process and being part of a group. When the breathing group is done, everyone sits around in a circle to share their experiences, which is known as group process. This is important because the integration of the experience can leave one very vulnerable to the outside world and this helps a person integrate the experience and helps with grounding one's self. Research suggests that humans are hardwire for group connection and that the human mind has developed as a social organ. The group creates a container for one to go through an experience together, which can help promote bonding.

Sitting: Breathing is done in pairs, in which one participant breathes while the other participant sits and provides support for the breathing. Sitting can be an important part of the breathing experience because it allows individual attention and provides a sense of safety to the breather. The sitter's job is to remain non-directive, provide support when needed, such as assisted the breather to the bathroom or hand the breather tissues.

Expressive Art: Lastly, everyone who breathes is asked to do a mandala or creative art piece. This is done to help integrate their powerful experience but also help with group processing. Engaging in expressive art after a breathwork session allows one to non-verbally express his or her experience. It is important to remind the participant to listen to his or her body and try not to overthink the process. This is an opportunity to allow the body to speak on paper. Facilitators encourage participants to hang his or her art piece where they can see it and reflect on the piece of art over time. The mandala can be an important part of one's integration process.

Discussion

Considering that dropout rates in therapy are as high as 54% for clients who suffer from PTSD (Seppälä et al., 2014), it is important to explore a variety of treatment option for clients who do not respond to the traditional therapies available. Seppälä et al. (2014) and others state that breathing practices can be beneficial for clients suffering from PTSD as breathwork helps to regulate the nervous system and calm overactive sympathetic responses. Since there is a lack of treatment success within traditional interventions for PTSD (Seppälä et al., 2014), breathwork should be examined for its efficacy, as it is a relatively safe modality and could have large implications in mental health treatment.

It is important to understand how each breathing practice is utilized and when each practice may or may not be appropriate since there are various fields of breathwork. When incorporating breathwork practices in a clinical setting, it is crucial to understand where the client is at emotionally, mentally, physically, and how he or she might respond to the interventions. After reviewing the breathwork literature, it appears that there is not a system of working with breathwork in a counseling setting that incorporates the various applications. Each field seems separate and not integrated. This paper is an attempt to synthesize a framework/system to work with breathwork as an intervention to trauma and PTSD in a clinical system.

There are limitations to the various breathing practices, which is why it is important to have a multilevel system of working with breathwork that it covers the different manifestations of trauma. This system proposes that breathwork can help regulate the nervous system and hyperarousal/hypoarousal states, help to resolve somatic manifestations of trauma, and deal with the biographical, transpersonal, and perinatal roots of trauma. After examining the trauma literature, it is safe to conclude that trauma does not just affect the mind, but it also affects the body and spirit of an individual. A more integrative approach is needed when it comes to treating trauma and PTSD.

There are some obvious limitations to this system as it has not been tested or studied. The proposed framework outlines what an integrative approach could look like. This paper does not suggest that this system is a *treatment* for PTSD and trauma, but rather an examination of what an integrative breathwork intervention could potentially look like. This paper's goal is also to highlight that breathwork research as an intervention for PTSD is limited and to stress the importance of breathwork research within the counseling field. Breathwork within the counseling field seems to exist in the "alternative" models, but these practices should be researched and examined for its potential as an evidence-based practice.

Finally, this paper's goal is to highlight the important potential clinical application of breathwork and to encourage further breathwork research within the field of counseling. One way to encourage research is through education. The bulk of this paper is the basis of a training or educational program that will be oriented toward counselors and explore the intersection of breathwork and trauma. Conducting research and filling the gap between counseling and breathwork may lead to new adjunct therapies that could be more effective than traditional top-down therapy approaches.

Conclusion

Trauma can profoundly impact anyone's life at any moment. Trauma and PTSD do not just affect the individual suffering but has a large impact on families, friends, and the larger community. It is important to explore and advocate for research regarding alternative interventions. An individual who suffers from PTSD should have various treatment options, as traditional treatments are limiting at times. Breathwork practices can be an effective intervention for treating trauma and PTSD and should be examined for its efficacy. When examining breathwork as an intervention for trauma disorders, it is also important to stress the importance of an integrative model that focuses on healing the mind, body, and spirit. The various systems of breathing can be utilized to help with emotional regulation, somatic processing,

Breathwork and Trauma Resolution

and foster self-discovery to cultivate a fuller and richer life experience.

References

Agorastos, A., Boel, J. A., Heppner, P. S., Hager, T., Moeller-Bertram, T., Haji, U., Motazedi, A., ... Stiedl, O. (May 01, 2013). Diminished vagal activity and blunted diurnal variation of heart rate dynamics in posttraumatic stress disorder. *Stress, 16,* 3, 300-310.

Akdemir, B. (2016). Vagus nerve stimulation: An evolving adjunctive treatment for cardiac disease. *The Anatolian Journal of Cardiology.*
https://doi.org/10.14744/AnatolJCardiol.2016.7129

American Counseling Association. (2011). *Fact Sheet #7* [PDF]. American Counseling Association. Retrieved from https://www.counseling.org/docs/trauma-disaster/fact-sheet-7---terms-to-know.pdf?sfvrsn=af3e0017_2

American Counseling Association (2014). ACA Code of Ethics. Alexandria, VA: Author

American Psychiatric Association. (2013). Diagnostic and statistical manual of mental disorders (5th ed.). Arlington, VA: Author.

American Psychological Association. (2018). *Trauma and Shock.* [online] Available at: https://www.apa.org/topics/trauma/.

American Psychiatric Association. (2018). *What Is PTSD?.* [online] Available at: https://www.psychiatry.org/patients-families/ptsd/what-is-ptsd [Accessed 16 Dec. 2018].

Aposhyan, S. (2004). *Body-mind psychotherapy,* New York, N.Y.: W.W. Norton & Company.

Bloom, S. L. (October, 1999). Trauma theory abbreviated. From *Final action plan: a coordinated community-based response to family violence.* [Pdf].

Bray, P. (2013). Bereavement and transformation: A Psycho-spiritual and post-traumatic growth perspective. *Journal of Religion &*

Health, 52(3), 890–903. https://doi-org.prescottcollege.idm.oclc.org/10.1007/s10943-011-9539-8

Brewerton, T. D., Eyerman, J. E., Cappetta, P., & Mithoefer, M. C. (June 01, 2012). Long-term abstinence following Holotropic Breathwork as adjunctive treatment of substance use disorders and related psychiatric comorbidity. *International Journal of Mental Health and Addiction, 10,* 3, 453-459.

Brown, R. P., & Gerbarg, P. L. (2012). *The healing power of the breath: Simple techniques to reduce stress and anxiety, enhance concentration, and balance your emotions.* Boston, Mass: Shambhala.

Caldwell, C. (1997). *Getting in touch: The guide to new body-centered therapies.* Wheaton, IL: Quest

Caldwell, C., & Victoria, H. K. (2011). Breathwork in body psychotherapy: Towards a more unified theory and practice. *Body, Movement and Dance in Psychotherapy, 6*(2), 89-101.

Cortina, M., & Liotti, G. (2007). New approaches to understanding unconscious processes: Implicit and explicit memory systems. *International Forum of Psychoanalysis, 16*(4), 204–212. https://doi.org/10.1080/08037060701676326

Cozolino, L. (2014). *The neuroscience of human relationships: attachment and the developing social brain.* 2nd Ed. New York, NY: W.W. Norton & Company.

Crawford, A. (2010). If "The Body Keeps the Score": Mapping the dissociated body in trauma narrative, intervention, and theory. *University of Toronto Quarterly, 79*(2), 702–719. https://doi-org.prescottcollege.idm.oclc.org/10.3138/UTQ.79.2.702

Crockett, J. E., Cashwell, C. S., Tangen, J. L., Hall, K. H., & Young, J. S. (2016). Breathing characteristics and symptoms of psychological distress: An exploratory study. *Counseling and Values, 61*(1), 10–27. https://doi.org/10.1002/cvj.12023

Descilo, T., Vedamurtachar, A., Gerbarg, P. L., Nagaraja, D., Gangadhar, B. N., Damodaran, B., ... Brown, R. P. (2010). Effects of a yoga breath intervention alone and in combination with an exposure therapy for post-traumatic stress disorder and depression in survivors of the 2004 South-East Asia tsunami. *Acta Psychiatrica Scandinavica*, *121*(4), 289–300. https://doi.org/10.1111/j.1600-0447.2009.01466.x

Egan, R. P., Hill, K. E., & Foti, D. (2018). Differential effects of state and trait mindfulness on the late positive potential. *Emotion*, *18*(8), 1128–1141. https://doi-org.prescottcollege.idm.oclc.org/10.1037/emo0000383.supp (Supplemental)

Eyerman, J. (2013). A clinical report of Holotropic Breathwork in 11.000 psychiatric inpatients in a community hospital setting. *Multidisciplinary Association for Psychedelic Studies Bulletin Special Edition*, *23*(1), 24–27.

Feduccia, A. A., & Mithoefer, M. C. (2018). MDMA-assisted psychotherapy for PTSD: Are memory reconsolidation and fear extinction underlying mechanisms? *Progress in Neuro-Psychopharmacology and Biological Psychiatry*, *84*, 221–228. https://doi.org/10.1016/j.pnpbp.2018.03.003

Felitti, V. J., Anda, R. F., Nordenberg, D., Williamson, D. F., Spitz, A. M., Edwards, V., ... Marks, J. S. (1998). Relationship of Childhood Abuse and Household Dysfunction to Many of the Leading Causes of Death in Adults. *American Journal of Preventive Medicine*, *14*(4), 245–258. https://doi.org/10.1016/S0749-3797(98)00017-8

Garland, E. L., & Howard, M. O. (2009). Neuroplasticity, psychosocial genomics, and the biopsychosocial paradigm in the 21st century. *Health & social work*, *34*(3), 191-9.

Grof, S. (2003). Implications of modern consciousness research for psychology: Holotropic experiences and their healing and heuristic potential. *The Humanistic Psychologist, 31*(2–3), 50–85. https://doi.org/10.1080/08873267.2003.9986926

Grof, S., & Grof, C. (2010). *Holotropic Breathwork: A new approach to self-exploration and therapy*. Albany: State University of New York Press.

Grof Transpersonal Training. (2018). *About Holotropic Breathwork*. [online] Available at: http://www.holotropic.com/holotropic-breathwork/about-holotropic-breathwork/ [Accessed 28 Dec. 2018].

Gupta, M. A. (2013). Review of somatic symptoms in post-traumatic stress disorder. *International Review of Psychiatry*, 25(1), 86–99. https://doiorg.prescottcollege.idm.oclc.org/10.3109/09540261.2012.736367

Haller, M., Myers, U. S., McKnight, A., Angkaw, A. C., & Norman, S. B. (2016). Predicting engagement in psychotherapy, pharmacotherapy, or both psychotherapy and pharmacotherapy among returning veterans seeking PTSD treatment. *Psychological Services*, *13*(4), 341–348. https://doi-org.prescottcollege.idm.oclc.org/10.1037/ser0000093

Hanson, R., & Mendius, R. (2009). *Buddha's brain.* Oakland, CA: New Harbinger Publications.

Holmes, S. W., Morris, R., Clance, P. R., & Putney, R. T. (1996). Holotropic Breathwork: An experiential approach to psychotherapy. *Psychotherapy: Theory, Research, Practice, Training, 33*(1), 114–120. https://doi.org/10.1037/0033-3204.33.1.114

Jerath, R., Crawford, M., Barnes, V., & Harden, K. (2015). Self-regulation of breathing as a primary treatment for anxiety. *Applied Psychophysiology & Biofeedback*, *40*(2), 107–115. https://doi-org.prescottcollege.idm.oclc.org/10.1007/s10484-015-9279-8

[joy96815]. (2013). *Stephen Porges "The Polyvagal Theory."* [Video file]. Retrieved from https://www.youtube.com/watch?v=8tz146HQotY

Kim, S., Roth, W. T., & Wollburg, E. (2015). Effects of therapeutic relationship, expectancy, and credibility in breathing therapies for anxiety. *Bulletin of the Menninger Clinic, 79*(2), 116–130. https://doi.org/10.1521/bumc.2015.79.2.116

Koven, & Steven G. (2018.) Veteran Treatments: PTSD Interventions. *Healthcare* 6, no. 3: 94.

Lalande, L., Bambling, M., King, R., & Lowe, R. (2012). Breathwork: An additional treatment option for depression and anxiety? Journal of Contemporary Psychotherapy, 42(2), 113-119.

Levine, P. (2010). *In an unspoken voice: How the body releases trauma and restores goodness.* Berkeley, CA: North Atlantic Books

Levenson, J. (2017). Trauma-informed social work practice. *Social Work, 62*(2), 105–113. https://doi-org.prescottcollege.idm.oclc.org/10.1093/sw/swx001

Li, Y., Hou, X., Wei, D., Du, X., Zhang, Q., Liu, G., & Qiu, J. (2017). Long-term effects of acute stress on the prefrontal-limbic system in the healthy adult. *PLoS ONE, 12*(1), 1–16. https://doi-org.prescottcollege.idm.oclc.org/10.1371/journal.pone.0168315

Lowen, A. (1975). *Biogeneretics: The revolutionary therapy that uses the language of the body to heal the problems of the mind.* New York: Penguin Group (USA).

Masten, C. L., Guyer, A. E., Hodgdon, H. B., McClure, E. B., Charney, D. S., Ernst, M., ... Monk, C. S. (2008). Recognition of facial emotions among maltreated children with high rates of post-traumatic stress disorder. *Child Abuse & Neglect, 32*(1), 139–153. https://doi.org/10.1016/j.chiabu.2007.09.006

Meg-rottweil. (2016). *Stephen Porges - Polyvagal Theory: how your body makes the decision.* Retrieved from https://www.youtube.com/watch?v=ivLEAlhBHPM

Mental Health America. (n.d.). Fact sheet: Understanding, preventing and healing trauma. [online] Available at: https://www.mentalhealthamerica.net/sites/default/files/Trauma_fact_sheet_final1.doc

Mithoefer, M. (2003). The physiology of hyperventilation. In: T. Kylea, ed., *Exploring Holotropic Breathwork: Selected articles from a decade of The Inner Door.* Santa Cruz, CA: Hanford Mead, pp.52-58.

Mithoefer, Michael, (January, 2017). A manual for MDMA-assisted psychotherapy in the treatment of Posttraumatic Stress Disorder. MAPS: Santa Cruz, CA.

Morningstar, J. (2017). *Break through with breathwork: Jump-starting personal growth in counseling and the healing arts.* Berkeley, California: North Atlantic Books

Niles, B. L., Polizzi, C. P., Voelkel, E., Weinstein, E. S., Smidt, K., & Fisher, L. M. (2018). Initiation, dropout, and outcome from evidence-based psychotherapies in a VA PTSD outpatient clinic. *Psychological Services*, *15*(4), 496–502. https://doi-org.prescottcollege.idm.oclc.org/10.1037/ser0000175

Nivethitha, L., Mooventhan, A., & Manjunath, N. K. (2016). Effects of various prāṇāyāma on cardiovascular and autonomic variables. *Ancient Science of Life*, *36*(2), 72–77. https://doi-org.prescottcollege.idm.oclc.org/10.4103/asl.ASLpass:[_]178_16

Porges, S. W. (1998). Love: An emergent property of the mammalian autonomic nervous system. *Psychoneuroendocrinology*, *23*(8), 837–861.

Porges, S. W. (2003). The Polyvagal Theory: phylogenetic contributions to social behavior. *Physiology & Behavior*, *79*(3), 503–513. https://doi.org/10.1016/S0031-9384(03)00156-2

Porges, S. W. (2007). The polyvagal perspective. *Biological Psychology, 74*(2), 116–143.

Reisman M. (2016). PTSD treatment for veterans: What's working, what's new, and what's next. *P & T: A Peer-Reviewed Journal for Formulary Management, 41*(10), 623-634.

Rhinewine, J. P., & Williams, O. J. (2007). Holotropic Breathwork: The Potential Role of a Prolonged, Voluntary Hyperventilation Procedure as an Adjunct to Psychotherapy. *Journal of Alternative & Complementary Medicine, 13*(7), 771–776. https://doi-org.prescottcollege.idm.oclc.org/10.1089/acm.2006.6203

Roncada, G., Vandevelde, B. and Calsius, J. (2018). The human body and psychological trauma: biological explanatory models. *International Body Psychotherapy Journal: The Art and Science of Somatic Praxis*, [online] 17(1), pp.22-50. Available at: https://www.ibpj.org/issues/IBPJ-Volume-17-Number-1-2018.pdf [Accessed 14 Nov. 2018].

Rothschild, B. (2017). *The body remembers volume 2: Revolutionizing trauma treatment.* New York: W.W. Norton & Company.

SAMSA-HRSA. (n.d.). Trauma. Retrieved from https://www.integration.samhsa.gov/clinical-practice/trauma-informed

Sapolsky, R. (2017.). *The biology of our best and worst selves.* Retrieved from https://www.ted.com/talks/robert_sapolsky_the_biology_of_our_best_and_worst_selves

Seo, D., Rabinowitz, A. G., Douglas, R. J., & Sinha, R. (2019). Limbic response to stress linking life trauma and hypothalamus-pituitary-adrenal axis function. *Psychoneuroendocrinology, 99*, 38–46. https://doi.org/10.1016/j.psyneuen.2018.08.023

Seppälä, E. M., Nitschke, J. B., Tudorascu, D. L., Hayes, A., Goldstein, M. R., Nguyen, D. T. H., … Davidson, R. J. (2014). Breathing-based meditation decreases posttraumatic stress disorder symptoms in U.S. military veterans: A randomized controlled longitudinal study. *Journal of Traumatic Stress*, *27*(4), 397–405. https://doi.org/10.1002/jts.21936

Sessa, B. (2017). MDMA and PTSD treatment: "PTSD: From novel pathophysiology to innovative therapeutics." *Neuroscience Letters*, *649*, 176–180. https://doi.org/10.1016/j.neulet.2016.07.004

Siegel, D. (2011). *Mindsight: The new science of personal transformation*. New York, NY: Bantam Books.

Siegel, D. J. (2015). *The developing mind, second edition: How relationships and the brain interact to shape who we are*. New York, NY, United States: Guilford Publications.

Stevens, J. (2017). *Addiction doc says: It's not the drugs. It's the ACEs...adverse childhood experiences*. [online] ACEs Too High. Available at: https://acestoohigh.com/2017/05/02/addiction-doc-says-stop-chasing-the-drug-focus-on-aces-people-can-recover/ [Accessed 2 Oct. 2018].

Substance Abuse and Mental Health Services Administration. (2018). *Trauma and violence*. [online] Available at: https://www.samhsa.gov/trauma-violence.

Taylor, K. (2003). *Exploring Holotropic Breathwork: Selected articles from a decade of The Inner Door*. Santa Cruz, Calif: Hanford Mead.

Van der Kolk, B. A. (2014). *The body keeps the score: Brain, mind, and body in the healing of trauma*. New York: Viking.

Victoria, H. K., & Caldwell, C. (2013). Breathwork in body psychotherapy: Clinical applications. *Body, Movement and Dance in Psychotherapy*, *8*(4), 216–228. https://doi.org/10.1080/17432979.2013.828657

Williamson, J. B., Porges, E. C., Lamb, D. G., & Porges, S. W. (2015). Maladaptive autonomic regulation in PTSD accelerates physiological aging. *Frontiers in Psychology, 5,* 1571. https://doi.org/10.3389/fpsyg.2014.01571

Young, J. S., Cashwell, C.S., & Giordano, A.L. (2010). Breathwork as a therapeutic modality: An overview for counselors. *Counseling and Values,* 55(1), 113-125.

Zalaquett, C. (2013). *Post Traumatic Stress Disorder* [PDF]. American Counseling Association. Retrieved from https://www.counseling.org/docs/default-source/practice-briefs/post-traumatic-stress-disorder.pdf?sfvrsn=ac6ff615_15

Appendix A

Glossary

Amygdala: A part of the limbic system in the brain that is responsible for emotional regulation and for processing emotions like fear.

Autonomic Nervous System (ANS): This system is associated with self-regulation and adaptability. The ANS is derived of the parasympathetic and sympathetic nervous systems. The ANS plays a crucial role in all physiological and behavioral functions that rely on cardiac, smooth, and striated muscles.

Breathwork: A system of conscious breathing techniques that foster healing, self-exploration, and self-discovery. Breathwork can include various breathing practices such as yogic or pranayama breathwork, Transpersonal Breathwork, and Holotropic Breathwork. These practices usually consist of controlled breathing patterns.

Dreamshadow Transpersonal Breathwork: A breathing technique that has roots in Stan and Christina Grof's Holotropic Breathwork that is taught by Elizabeth and Lenny Gibson.

Holotropic Breathwork: A breathing technique that was developed by Stan and Christina Grof, which consists of deep circular breathing, evocative music, focused body work, and group process.

Parasympathetic Nervous System (PNS): The calming system that helps to dampen the excitatory nervous system and is associated with rest, energy conservation, digestion, and lowering the heart rate.

Posttraumatic Stress Disorder (PTSD): A psychiatric/mental health disorder that occurs in people who have witnessed or experience a traumatic event, which causing prolonged negative symptoms that can be debilitating such as flashbacks, anxiety, nightmares, intrusive/negative thoughts, depression, and avoidant behaviors.

Smart Vagus: A branch of the ventral vagus nerve that is known as the social engagement system. It is crucial for emotional attunement, caregiving, and sustained social contact.

Sympathetic Nervous System (SNS): The excitatory system that is responsible for increasing energy for fight or flight behaviors and impacts heart rate, sweat glands, and blood pressure.

Trauma: The American Psychological Association (2018) defines trauma as "an emotional response to a terrible event like an accident, rape or natural disaster." The emotional response can manifest as distress, anxiety, unpredictable emotions, flashbacks, or physical/somatic sensations.

Polyvagal Theory: A theory that was developed by Dr. Stephen Porges to explain how autonomic function relates to and influences behavior.

Vagus Nerve: The tenth cranial nerve that runs from the brainstem to multiple points in the body. The word *vagus* means, wandering. This nerve wanders down the brainstem to the throat, heart, lungs, and digestive system and is important for respiratory function, heart rate, digestion and social engagement/communication.

Acknowledgments

A special thank you to Lenny and Elizabeth Gibson, and Michael Watson and Jennie Kristle. I cannot express enough gratitude or love to you all for supporting me on my journey and passing down your teachings and wisdom.

Special thanks to those who have helped with editing this paper/book: Elizabeth Gibson. M.S., Alan Kooi Davis, Ph.D., Davida Taurek, M.S., and Janys Murphy, Ph.D.

And of course, my parents and family who have supported me along the way

Much love!

About the Author

Kyle Buller, M.S.

Kyle's interest in exploring non-ordinary states of consciousness began at the age of 16 when he suffered a traumatic snowboarding accident. After this near-death experience, Kyle's life changed dramatically. Kyle subsequently earned his B.A. in Transpersonal Psychology from Burlington College, where he focused on studying the healing potential of non-ordinary states of consciousness by exploring shamanism, Reiki, local medicinal plants and plant medicine, Holotropic Breathwork, and psychedelic psychotherapy. Kyle has been studying breathwork with Lenny and Elizabeth since October 2010.

Kyle earned his M.S. in clinical mental health counseling with an emphasis in somatic psychology from Prescott College. Kyle's clinical background in mental health consists of working with at-risk teenagers in crisis and with individuals experiencing an early-episode of psychosis and providing counseling to undergraduate/graduate students in a university setting. Kyle also facilitates Transpersonal Breathwork workshops.

Learning more about Kyle at:

www.settingsunwellness.com

www.psychedelicstoday.com

www.ingramcontent.com/pod-product-compliance
Lightning Source LLC
Chambersburg PA
CBHW021309240526
45463CB00018B/797